SUCCESSFUL PROPOSAL STRATEGIES ON-THE-GO!

For a complete listing of titles in the
Artech House Technology Management and Professional Development Library,
turn to the back of this book.

SUCCESSFUL PROPOSAL STRATEGIES ON-THE-GO!

Robert S. Frey

ARTECH HOUSE
BOSTON | LONDON
artechhouse.com

Library of Congress Cataloging-in-Publication Data
A catalog record for this book is available from the U.S. Library of Congress.

British Library Cataloguing in Publication Data
A catalogue record for this book is available from the British Library.

Cover design by Joi Garron

ISBN 13: 978-1-68569-013-7

Artech House
685 Canton Street
Norwood, MA 02062

10 9 8 7 6 5 4 3 2 1

To my longtime friend and colleague, Dave Santoro—
quality management guru, whose 8-decade journey through life
has left him cheerful, insightful, and perennially positive

And to the late H. Moffette Tharpe, Jr. (1939–2022),
who was one of my primary mentors in the federal
government proposal marketspace

CONTENTS

FOREWORD *xv*

PREFACE *xvii*

ACKNOWLEDGMENTS *xix*

1
FOCUSED PASSION 1

1.1 Leadership Across Time and Geography 1
1.2 Power of Proposal Win Bonuses 3
1.3 Proposals and Proposaling in Context 6
1.4 Focused Passion Helps Win Federal Proposals 8
1.5 First Things First 9
1.6 Inspiring Passion 11
References 12

2
PROPOSALING IN CONTEXT: WHERE DOES THE PROPOSAL FIT WITHIN THE ENTIRE MARKETING LIFE CYCLE? 13

2.1 Introduction 13
2.2 Wrapping Business Development, Capture Management, and Proposal Development in a Blanket 14

2.3 Everyone in Your Company Is in Business Development 16
2.4 Looking Through the Lens of PMBOK 18
2.5 Using the Lens of PMBOK to Focus on Capture Management 20
2.6 Proposal Development—Pathway to Successful Blue Team Review 22
2.7 KM Through the Lens of PMBOK 24
References 26

3

THE IMPORTANCE OF HUMAN DYNAMICS 27

3.1 Metaphors Matter 27
3.2 The 71% Effect 30
3.3 Stop Drowning in Meetings 32
3.4 Searching for Effective Communication 33
3.5 Organizational Culture and Performance Success 34
Selected Bibliography for Building Effective Teams 36
References 37

4

PROPOSAL DEVELOPMENT AND KNOWLEDGE MANAGEMENT 39

4.1 Collaboration Tools 39
4.2 Spinning Proposal Gold 41
4.3 Agile Project Management and Effective Knowledge Sharing in Organizations 43
4.4 Lighting the Way—Getting the Most Out of What You Know 45
4.5 Artificial Intelligence and KM in Proposal Land 47
References 49

5

STATED AND UNSTATED CRITERIA 51

5.1 Stated and Unstated Criteria 51
5.2 Deeper Insight into Unstated Evaluation Criteria 52
5.3 Key Questions to Ask Government Civil Servants to Learn About Unstated Criteria 53

5.4 Getting Credit for Your Strengths—Shaping the Final RFP 55
5.5 Candidate Instructions to Offerors Language (Section L) 56
5.6 Candidate Evaluation Factors Language (Section M) 56
5.7 Candidate Section L Language: Long-Term Business Strategy 57
5.8 Candidate Evaluation Factor Language: Long-Term Business Strategy 58
5.9 Candidate Section L Language: Process Innovations 58
5.10 Candidate Evaluation Factor Language: Process Innovations 59
5.11 45 Strengths Are the New Target 59
Reference 59

6

MOVING BEYOND COMPLIANCE 61

6.1 Proposal Compliance Is Necessary but Not Sufficient 61
6.2 Font and Line Spacing: Still Critical Aspects of Proposaling 63
6.3 Straightforward Compliance Can Be Challenging 64
6.4 Building a Meaningful Compliance Matrix 66

7

THE LOOK AND FEEL OF A WINNING PROPOSAL 71

7.1 What Does a Winning Proposal Look Like? 71
7.2 Good Is the Enemy of Great 73
7.3 The Most Important Factors in Winning 75
7.4 Staying Sharp—Business Acquisition as a Formal Process 78
References 80

8

HOPSCOTCHING THROUGH PROPOSAL-LAND 81

8.1 Proposal Integration Map 81
8.2 Framing and Proposal Development 83
8.3 Making Your Business Processes Stand Out 85
8.4 Communication as a Critical Success Factor in Proposal Development 86
8.5 Federal Proposal Development—Comprehensive Risk Management Approach 88

8.6 Federal Government Proposals—Responsible Contractor 89
8.7 Take Contractor Responsibility Determinations Seriously 90
8.8 Oral Presentations 91
8.8.1 Best Practices for Virtual Oral Presentations 91
8.8.2 Selecting the Optimal Oral Presenters 92
8.9 Inquiring Minds 94
References 98

9

THE IMPORTANCE OF UNDERSTANDING: SECTIONS L AND M 101

9.1 Paint a Picture of Genuine Understanding in Your Proposals 101
9.2 Missing the Point of Understanding 104
Reference 109

10

ORCHESTRATING TEAM INTERACTION AND DEVELOPING AN ON-TARGET PROPOSAL APPROACH 111

10.1 Proposal Directive—Living Encyclopedia of a Win 111
10.2 The Critical Role of the Proposal Manager 113
10.3 Pulling Intriguing Threads—Interviews with Technical SMEs 116
10.4 Rules of Engagement for Effective Proposal Meetings 119
10.5 Building Your Technical Approach Sections for a U.S. Government Proposal 120
10.6 Increasing Effectiveness in Remote Proposal Development Environments 123

11

USING SECTION M AS A WINDOW 125

11.1 Federal Proposal Development—Looking Through the Window of Section M 125
References 130

12

PROPOSAL SOLUTION DEVELOPMENT 133

12.1 Solution Development Overview 133
12.2 Solution Development Tools 134
12.3 Solution Development Process 134
12.4 Framework for Technical Solution Development 136
12.5 Box-in-a-Box Model for Proposal Development 137
12.6 Solution Development—Effective Alternative to Brainstorming 139
12.7 Helpful Proposal Publications 141
Reference 146

13

PRECURSOR TO THE EXECUTIVE SUMMARY 147

13.1 Introduction 147
13.2 Win Strategy White Paper as a Solution Development Tool 149
13.3 The Executive Summary as a Requirement 152

14

MULTIPLE PALETTES 155

14.1 Another Palette: A Meaningful Proposal Cover Letter 155
14.2 Proposal Graphics and Proposal Cover Design 157
14.3 Websites—Proposal Support Platform and Source for Protest 162
14.4 Proposal Protest Source Materials 163
14.5 LinkedIn Profiles of Key Personnel 164
References 165

15

PROPOSAL WRITING 167

15.1 Proposals Are Knowledge-Based Sales Documents 167
15.2 ABCs of Proposal Writing 168
15.3 Writing Standards 170
15.4 Abstracting as a Proposal Skill 171
15.5 Active Voice Adds Strength and Saves Space 172
15.6 Action Captions 175

15.7 Methods of Enhancing Your Proposal Writing and Editing 176
15.8 Resources to Support Your Company's Proposal Writing Efforts 178
15.9 Potpourri 179
15.10 Take Off in the Right Direction: Defining Acronyms Early in Your Proposal Development Process 179
15.11 Dovetailing—Fit Your Proposal Exactingly with the Federal RFP 181
15.12 Words to Avoid in Your Proposals 183

16

ACHIEVING PROPOSAL STRENGTHS THROUGH FORWARD-LOOKING BUSINESS DECISIONS 187

16.1 Source Selection Documents and Proposal Strengths 187
16.2 Linking Key Business Decisions to Proposal Strengths 190
16.3 Pointing the Way—Answers to Training Participants' Questions 191
16.4 Post-Award Debriefings 191
16.5 Leveraging Source Selection Statements to Make Key Business Decisions 192

17

LOOKING AT PROPOSAL REVIEWS FROM DIFFERENT ANGLES 195

17.1 Vertical and Horizontal Proposal Reviews 195
17.2 Measure What Is Important—Ensuring That Strengths for Your Company Appear in the Government's Source Selection Statement 197
17.3 Combining Agile Scrum and SAM 200
17.4 Agile in Context 210
Selected Bibliography 211
References 212

18

BUILDING THE COST/PRICE VOLUME 215

18.1 Overview of the FAR 215
18.2 Defining Key Pricing Terminology 216
18.2.1 Cost Objective 216

18.2.2 Final Cost Objective 217
18.2.3 Direct Costs 217
18.2.4 Direct Labor 217
18.2.5 Subcontractor Labor 218
18.2.6 Labor Categories 218
18.2.7 Other Direct Costs 218
18.2.8 Indirect Costs 218
18.2.9 Fringe Benefits 219
18.2.10 Overhead 219
18.2.11 Procurement 220
18.2.12 G&A 220
18.2.13 Fully Burdened Rate 221
18.2.14 Fee 221
18.2.15 Unallowable Costs 221
18.3 Overview of Federal Travel Regulations 222
18.4 Overview of the Cost Accounting Standards 222
18.5 Overview of Relevant GAAP 223
18.6 Adequacy of the Contractor's Internal Accounting System 223
18.7 The DCAA 224
18.8 RFP Review 226
18.9 Importance of Aligning the Cost Proposal with the Technical and Management Proposals 226
18.10 Cost Estimating Methods 227
18.11 Methodology for Determining Salary Ranges 227
18.12 Methodology for Computing Labor Escalation 227
18.13 Highlights of the Process to Calculate Indirect Costs 228
18.14 Determining Annual Productive Labor Hours 229
18.15 Methodology to Compute Fee 229
18.16 Cost Volume Narrative and Production 229
18.17 Audit Files 230
18.18 Special Topics 231
18.18.1 Uncompensated Overtime 231
18.18.2 Addressing Risk in the Cost Volume 231
18.18.3 What Is Cost Realism? 232
18.18.4 What Is Cost Reasonableness? 232
18.18.5 What Is Cost or Pricing Data? 233

18.18.6 Price to Win 234
18.18.7 Common Cost Proposal Problems 234
18.18.8 Useful Websites (Note That They May Change) and Templates (Always Check for the Latest Template Online) 235
18.19 Price Proposal Reference 236

19

ACADEMIC AND GOVERNMENT GRANT PROPOSALS AND INTERNATIONAL AND PRIVATE-SECTOR PROPOSALS 241

19.1 Federal Grants 241
19.2 Federal Contracts 241
19.3 Comparison of Grant Proposal Writing and Competitive Federal Proposal Writing 242
19.3.1 Grant Proposal Writing 242
19.3.2 Competitive Federal Proposal Writing 243
19.4 International Commercial Proposal Development 244

20

EPILOGUE 247

20.1 Direct Benefits of Federal Subcontracting Goals, Strategic Partnering, and Mentor-Protégé Relationships 247
20.2 Specific Strategies for Achieving Federal and Private-Sector Subcontracts 249
20.3 Best-Practice B&P Scenario—Shift Left 251
20.4 So, You Are Just Getting Started with Proposals 253

LIST OF ACRONYMS 259

ABOUT THE AUTHOR 275

INDEX 277

FOREWORD

In early 1998, I joined a small, upstart company, RS Information System (RSIS), Inc., whose mission was IT and engineering services to the U.S. federal government. I was an equity partner and responsible for advancing the growth of the company. The two other principals were aggressive, passionate, and entrepreneurial, but were struggling to win competitive proposals. The year before I arrived, they had submitted 17 proposals and had lost all 17. It was obvious that organizational and process enhancements had to be made in business development (BD), capture management, and proposal development/knowledge management. My strengths were more oriented to BD and capture management, not so strong in proposal and knowledge management.

One of the best executive hires I have made in my career is Dr. Bob Frey. He is singularly the most creative and productive person I know in this profession. Bob and I were convinced from the start that we could turn this situation around... and we did! RSIS was a marketing-driven organization. We believed that "if you're not growing, you're dying." We attracted the right people for the three important functions, communicated clear expectations, and rewarded successful results. We understood that proposal evaluators buy from people they know and trust... and they buy emotionally. We would do our best to gain customer intimacy before the request for proposal (RFP) hit the street so our proposed solutions would resonate. Bob developed a robust and effective knowledge management approach that allowed us to improve quality and maintain it over time.

For the next 10 years, we grew dramatically. From a few employees to more than 2,000 at various facilities around the country. Sales grew to over $500 million per year. Importantly, we achieved this organically, not through acquisitions. RSIS was one of the fastest growing companies in the nation, making it into the *Washington Technology* Fast 50 an unprecedented 8 years in a row! During this time, we won well over 100 contracts/task orders. Our win rate exceeded 60% and we were never the lowest priced winner. This speaks to the quality and best value of our offers.

Quite by chance, Bob Frey launched his own professional consultancy right at the onset of the Great Recession of 2008–09, which was the worst economic downturn in the U.S. since the Great Depression. Sixteen years later, he has supported well over 100 unique customers. I've always admired the fact that Bob did not bring on partners or independent contractors. "That's never been a part of my business model," he tells me.

Dr. Frey is an inspirational leader and a very effective communicator, teacher, and coach. He knows the value of forming a well-organized, integrated, and motivated team for each proposal pursuit. Bob has conducted many proposal seminars for hundreds of industry and government personnel. His published book entitled, *Successful Proposal Strategies for Small Businesses,* is now in its sixth edition. The didactic nature of his writing comes through by showing real-life examples of winning approaches and solutions.

Through a series of carefully arranged and easy-to-access short stories, Bob's new book—*Successful Proposal Strategies: On-The-Go!,* captures the insights, lessons learned, and best practices he has gained and applied, particularly in the past 11 years since the release of the sixth edition of his proposal book. My recommendation is that you as a business owner, capture manager, or proposal professional take Bob's field-proven guidance to heart. You'll see the positive difference it will make.

Ron G. Trowbridge
Chief Executive Officer
Blackwatch International
October 2023

PREFACE

Proposals have been and continue to be a significant dimension of my life. When I worked on my first proposal in the summer of 1987, I did not understand the difference between a red team and a red light. But I found the proposal process to be engaging and invigorating. Fast forward 36 years and literally thousands of U.S. federal government, commercial, and international proposals and grants later, and I still find the dynamism of proposal development to be captivating. The pandemic interrupted my favorite part of proposaling—human interaction. During my career, I have built and sustained many personal friendships with people from organizations all over the United States.

There's no theory in this book. It's a compendium of up-to-date, real-world vignettes. There is meaningful context built around each vignette, allowing readers to see immediately how to apply the lessons learned and guidance within their own organizations. The insights that I present are applicable to small businesses and midtier companies, as well as global *Fortune* 50 corporations. Topics are drawn from actual challenges and situations that organizations and their staff professionals face on a typical proposal. Across the spectrum of vignette topics, I paid attention to multiple dimensions in and around proposal development—human and organizational dynamics, linking business decisions to proposal strengths, building the proposal response, proposal writing, important metrics, proposal reviews, and proposal integration. In addition, coverage extends to academic and public-sector grant proposals, as well as international private-sector tenders. All of the vignettes are easy to use and integrate into the thinking and best practices of an organization precisely because they are streamlined. Importantly, *On-the-Go!* brings practical value to executive leadership, business developers, capture managers, and proposal developers and managers, along with technical and programmatic subject matter experts/content providers and knowledge managers. Digestible increments make picking up and engaging right away with *On-the-Go!* quite easy.

In an article published by the *Forbes* Agency Council entitled, "8 Big Trends in Publishing in 2023,"[1] two of the eight trends focused on artificial intelligence (AI) and natural language processing (NLP). The Council cautions that "some of the new AI applications out there have a different level of authenticity than a real person and are not fact-checking." Let me assure you as my readers that every word of *Successful Proposal Strategies On-The-Go!* has been written by a human being.

Now go win more stuff!

1. https://www.forbes.com/sites/forbesagencycouncil/2023/02/03/8-big-trends-in-publishing-in-2023-and-how-marketers-can-leverage-them/?sh=cdf9e3c1882c.

ACKNOWLEDGMENTS

Three people provided insightful content to two different chapters of *Successful Proposal Strategies On-the-Go!* Mike Parkinson contributed to Section 14.2 of Chapter 14: Multiple Palettes. The author of the book, *Do-It-Yourself Billion Dollar Graphics,* now in its second printing, Mike is a sought-after keynote speaker at worldwide venues on such topics as persuasion science, PowerPoint for education or sales, visual solution architecture, and proposal design. I have known and worked with Mike and his company, 24 Hour Company, for more than 20 years.

Kevin McQuade built, and Matt McKelvey updated, all of the material in Chapter 18: Building the Cost/Price Volume. As the president and CEO of the highly successful MSM Group, Inc., in Fairview Park, Ohio, Kevin has given lectures on accounting and regulatory issues affecting federal government contractors and has taught classes on government contract pricing techniques. He and I have worked elbow-to-elbow on federal proposals, as I have done with Matt McKelvey. Matt serves as the president of the Maryland-based financial consulting firm, The McKelvey Group (TMG). His company provides services across finance and accounting, as well as pricing support for government contractors and customized training.

Once again, my longtime friend and colleague, Lisa Richard, generated all of the line art and formatted the manuscript for *Successful*

Proposal Strategies On-the-Go! She is a highly talented and incredibly responsive professional. Lisa provided similar expert documentation services for each of the six editions of *Successful Proposal Strategies for Small Businesses* (Artech House). She is the owner of Lisa's Offsite Typographical Services (LOTS) in Austin, Texas.

1

FOCUSED PASSION

As a professor of program management and the executive coach at the Defense Acquisition University's (DAU) Capital and Northeast Region at Fort Belvoir, Virginia, Brian Schultz has written about the seven traits of high-performing teams [1]. Trait number 6 is passion [1]: "Teams that possess passion do not accept the status quo and seek new approaches and ways to improve." Schultz suggested that genuine passion manifests itself in the energy and focus that people bring to the job. Importantly, it is the organizational culture that ensures that passion and the other six "traits can develop and flourish" [1]. Writing in *Harvard Business School Working Knowledge,* in 2023, Silverthorne suggested that managers should "lead *for* passion" and "design *for* passion" [2]. Citing research that Harvard Business School professor Dr. Jon M Jachimowicz conducted, Silverthorne recommended that managers should learn what employees are passionate about, and create an "environment that keeps passion alive" [2].

1.1 LEADERSHIP ACROSS TIME AND GEOGRAPHY

In late 2022, I finished working closely with a small team to develop and submit a final proposal revision (FPR) on a major civilian agency proposal. At that time, it had been exactly 1 year since I had begun my support for the original proposal submission. As is common with complex procurements, a multicompany team was built. Among the subcontractors were an international firm with nearly $2 billion in U.S. dollars in annual revenues and another corporation with annual revenues higher than $140 million in U.S. dollars. Leading the team was a joint venture (JV) composed of two small businesses, one located in the Midwest and the other in the South. And that is where the ordinariness ended.

A year in advance of release of the final request for proposal (RFP) for this particular opportunity, the two leaders of the JV had invested the time and energy to forge a professional working relationship that was mutually supportive. It was that solid business foundation, coupled with the highly respectful nature of each of these leaders, which sustained the entire team across time (see Figure 1.1) and geography.

I first met one of the leaders of the JV and representatives from across most members of the team at a 3-day face-to-face (F2F) working session in Florida. Each and every person was made to feel welcome. The leaders of the JV knew the value of the participants and their organizations and communicated that merit forthrightly. To help establish meaningful camaraderie, there was a team dinner held at a

Figure 1.1 Clock over the Baltimore Orioles scoreboard. (Photograph © Dr. R. S. Frey.)

local restaurant on one of the evenings. It is so helpful to get to know the people with whom one is working outside of a structured business environment.

As release of the final RFP approached, the JV convened a second F2F working meeting, this time in Alabama. Importantly, representatives from all subcontractors were present. A senior leader from the international firm invested his time to participate in the daily solutioning sessions. Each night, the team was hosted for dinner at a cross-section of fine local restaurants.

During the proposal response life cycle, the JV held regularly scheduled videoconference meetings. Once again, the two leaders of the JV were inclusive. They openly recognized the contributions of the subcontractors, as well as the consultants who were engaged to provide customer-specific knowledge, proposaling insights, cost strategizing, and proposal development support. Not a great deal of time was spent on this recognition, but it happened frequently and it was heartfelt.

Seven months after proposal submission, the JV was notified by the government that they had made the competitive range in accordance with Federal Acquisition Regulation (FAR) 15. Seamlessly, the team came back together to focus considerable time and energy on responding to the government's findings. Resolution of the findings and oral discussions were then followed with the FPR. Leadership from the JV shone most brightly during the 50 days that comprised the competitive range response period. Verbal and written support and recognition were clearly evident. Team members willingly joined Zoom or MS Teams meetings at 6:00 P.M. on a Sunday evening, and there were people in 2 time zones to accommodate.

Across more than 36 years of professional proposal development and literally thousands of proposals, I have never experienced and witnessed such supportive and kind leadership. It brought out the very best in everyone involved.

1.2 POWER OF PROPOSAL WIN BONUSES

For nine years as a full-time employee for a federal government support services contractor, I participated directly in a highly successful proposal win bonus program. The operative word here is "win."

Working weekends and holidays, as well as long nights, on an unsuccessful proposal effort produced no bonus or recognition. The company had to be the prime and be the winner.

The bonus program was introduced to staff during the interview process. My formal offer letter in early 1999 referenced the program. With the exception of the four owners and the in-house business development staff who were compensated and rewarded according to a different framework, everyone in the company was bonus-eligible. That included all direct billable and overhead staff—in effect, operations, human resources (HR), recruiting, contracts and accounting, proposal development, quality assurance, and administrative support. The corporate vice president for business development managed and facilitated the program, with input from division vice presidents and other managers with direct knowledge of the proposal at hand.

Amazingly, within 30 days of award announcement by the government (not the contract start date), the Business Development Department organized an off-site, after-hours Proposal Win Party. Venues included local hotel conference rooms. In preparation for the Win Party (see Figure 1.2), accounting cut checks for each person who was to be recognized for his or her contribution to the proposal based upon guidance from the corporate vice president for business development. The company used a three-tier framework. The available bonus pool for a given proposal win was based upon total contract value

Figure 1.2 Red fireworks over Washington, D.C. (Photograph © Dr. R. S. Frey.)

(TCV) or strategic importance and was capped by the owners. In the case of indefinite delivery/indefinite quantity (ID/IQ) opportunities that carried no real dollar value, bonuses were still paid. That was because the ID/IQs could be monetized by winning task order proposals. In many cases, the ID/IQ, government-wide acquisition contract (GWAC), or multiple award contract (MAC) represented the only contractual gateway through which the company could pursue work in a given federal agency as the prime contractor. When small but geographically significant contracts were won, again, proposal win bonuses were still paid. For example, when the company won its first contract west of the Mississippi River, that geographic penetration was recognized and rewarded with bonus dollars.

Individuals who had made substantial contributions to the proposal in terms of extra time invested or exceptional technical insight, for example, were in Tier 1 (the highest level). Recruiting and HR members who orchestrated and staffed multiple open houses or technical interchange meetings (TIMs) across the country might also be Tier 1 recipients. Individualized certificates were printed with the name of the winning proposal effort and were signed by members of the executive team. Catered food and beverage services were arranged through the party venue.

At the beginning of each Win Party, the corporate vice president for business development along with members of the executive team, select business developers who were affiliated with the particular pursuit, and the cognizant division vice president responsible for performing the work gave enthusiastic opening remarks. In turn, every contributor to the proposal win was called to the front of the room and recognized publicly by their peers and management staff. Many win bonus recipients framed and displayed their certificates in their offices or cubicles. The names of all recipients were listed in the company's newsletter, as well as shown on slides during all-hands meetings. In this way, staff working at remote locations also received recognition for their contributions to winning.

Importantly, there was no limit to the number of proposal win bonuses that an employee could receive in a given year. As long as the individual was still employed by the company at the time of the Win Party, he or she reaped the rewards in proportion to the level of their contribution.

Now, I have also experienced situations wherein my employer offered no proposal bonuses of any kind. Then there is the group pizza lunch with the people with whom you just worked intensively during the entire proposal response life cycle. How uplifting! At another company, I distinctly remember that my proposal development team—a group of eight people including myself—received a single fruit basket for contributing significantly to a major win with the United States Agency for International Development (USAID). This was a new customer for this particular company. The apples and pears were tasty, but only went so far.

The great thing about the Proposal Win Bonus Program described here was that the company knew that the bookings were there when the bonuses were paid. In effect, the corporate bank account increased with the win. This type of program is critical to long-term business success. Especially now, when experienced proposal staff are in high demand and short supply. Lack of recognition results in career changes due to burnout.

Bottom line, businesses should tangibly encourage their staff to want to invest their time, energy, and passion to help win new and recompete work. It only makes sense for the bottom line.

1.3 PROPOSALS AND PROPOSALING IN CONTEXT

Winning. There are few feelings in the professional world of competitive federal government contracting that are sweeter or more fleeting. So how do you achieve this feeling in the first place, when there are other really talented and competent individuals in this same marketspace, seeking this feeling for themselves on the same program or project on which you have your own eyes set. Then the challenge becomes how to sustain winning over time, when the contract you hold dear comes up for recompetition after 5 years, or whatever the period of performance happens to be. By the way, this same feeling results from winning private-sector and international proposals, as well as grant proposals. The overwhelming majority of insights, lessons learned, and best practices that I have gained through 3.5 decades of experience in the federal space are applicable to these other proposal arenas as well, including state and local government opportunities.

Successful proposaling is as much art as it is science. In fact, probably more so. The art involves a level of nuance that far exceeds paint-by-number or sculpt-by-number approaches related to crafting a winning proposal. Such as explicitly addressing all of the "shall statements" found in the statement of work (SOW). Except that, the SOW or performance work statement (PWS) spans 245 pages found in Attachment J, not Section C where we are taught to find it, and you, Mr. or Ms. Contractor—or offeror, in proposal parlance—are allowed a mere 25 pages to present your detailed understanding and approach. What to do? Cloud-based templates are not the answer, certainly not on a consistent basis.

In the world of professional baseball, 20 Gold Gloves are normally awarded annually (see Figure 1.3). That represents 2% of the approximately 1,000 Major League players across 30 teams. Sparkling fielding displayed over 162 games is not based upon mechanics alone. Neither is addressing proposal-related challenges.

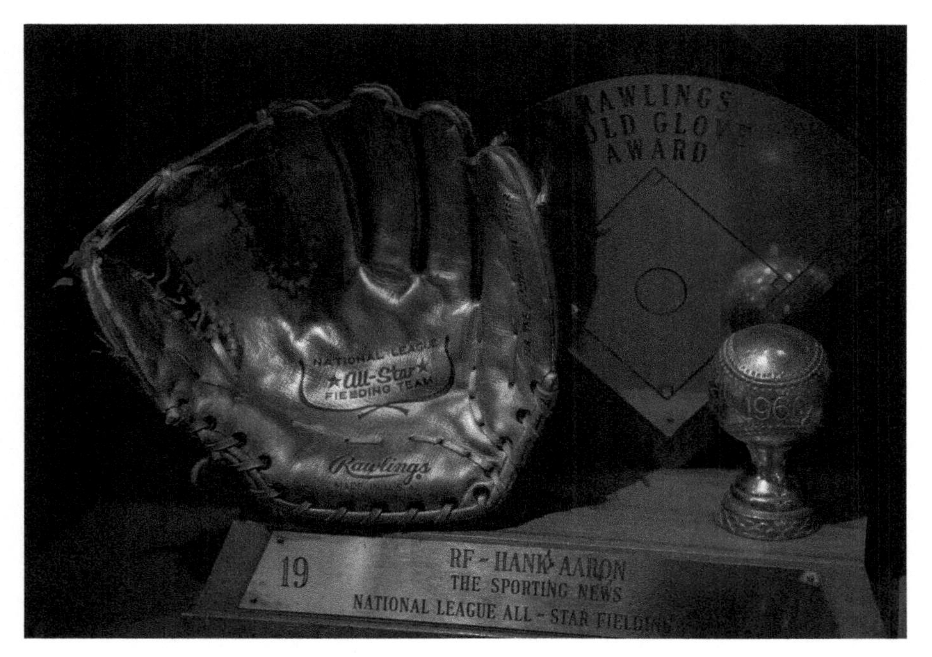

Figure 1.3 Gold Glove award to Hank Aaron. (Photograph © Dr. R. S. Frey.)

1.4 FOCUSED PASSION HELPS WIN FEDERAL PROPOSALS

Focused passion stands as a critical success factor in winning federal government proposals on a sustained basis. Right off the bat during the kickoff meeting (see Figure 1.4), emphasize teambuilding and mutual support. It sounds basic, but introduce the people individually and the areas in which they will contribute. Over time, introduce any new people added to the team (e.g., subject matter experts (SMEs) and reviewers). Communicate their respective roles and their counterparts across the entire team—prime contractor and subcontractors.

Moving forward throughout the entire proposal development life cycle, sustain a high level of esprit de corps. For example, do not allow outbriefings following review cycles to devolve into the negative—something I have witnessed far too frequently. Encourage and recognize achievements, significant contributions, and successes along the way. This is where executive leadership's direct and personalized engagement is critical. Engaged executive leadership drives the business culture of an organization to focus on vital activities, such as optimizing, encouraging, and leveraging human talent, continuous

Figure 1.4 Red, white, and blue baseball bats. (Photograph © Dr. R. S. Frey.)

learning and openness to change, trust-based knowledge sharing, and situational awareness across the competitive landscape.

Late one night after a continuous series of long nights, while the team I was supporting was working on a $92 million NASA Goddard Space Flight Center (GSFC) engineering support services proposal, the owner came into the office—in his elegant night clothes, no less—and gave everyone a crisp new $100 bill. Talk about an energizer! And yes, the proposal was a winner.

1.5 FIRST THINGS FIRST

The following are real questions that emerged during a multiday, F2F proposal training seminar that I conducted in Huntsville, Alabama. My answers (A) follow each question (Q).

Q: The DRAFT RFP (DRFP) was just released. Name the five most important things that we must do during the Week 1 post-DRFP.

A: 1. Conduct an objective gap analysis across the entire DRFP. Socialize the results internally within your company. This drives teaming decisions. Implement the service level agreement (SLA)-driven Attachment A for teaming agreements. Candidate language for Teaming Agreement, Attachment A is:

- The team member shall support the prime's[1] comarketing efforts for the entire duration of the pre-FINAL RFP timeframe, including providing specific SMEs to participate in marketing strategy sessions and at select F2F customer meetings.
- In accordance with workshare percentage or full-time equivalents (FTEs), the team member shall provide financial cost-sharing support for the Project ABC.
- During the proposal process, the team member shall make one specific person the primary point of contact for easy and direct access by the prime. That person or his or her designee shall also be available for direct, F2F participation in strategy and solution development sessions, formal and informal re-

1. The prime contractor works directly with the government. They manage any subcontractors and are responsible for ensuring that the work is completed as defined in the contract.

view meetings, oral presentation development meetings, and evaluation notice (EN)/final proposal revision (FPR) meetings.

- The team member shall acknowledge all preproposal, proposal, and post-proposal data calls from the prime within 4 hours of the time that the request is submitted.
- Specific technical, management, past performance, résumés, and pricing work products must be delivered to the prime in a timely and complete manner in accordance with the prime's master proposal schedule, formatting guidance, and page count targets.

2. Develop and populate the "Section M-to-Evidence of Strengths table" with quantitatively rich strengths. Section M of a federal RFP for an agency that follows the FAR is the Evaluation Factors for Award. Align validated strengths that your company is offering with those evaluation factors.
3. Develop a graphical depiction of your team's overall technical approach. I call this type of illustration a centerpiece graphic.
4. Develop a graphical depiction of the prime's overall management approach. Be sure to show interfaces with the government agency. By the way, in proposals, I always capitalize the word "government," as a gesture of respect to the entity that will be signing the checks.
5. Verify that you and your team have recent and referenceable past performance citations that have outstanding Contractor Performance Assessment Reporting System (CPARS) reports.

Q: Out of all you have conveyed, what should be our company's top 3 priorities regarding our business development (BD) and proposal development processes?

A: 1. Document and communicate the BD-capture-proposal process that your company will use going forward. The results of this practice also will prove valuable as a training tool when onboarding new staff professionals.
2. Know what you know—focus on knowledge management (KM). Knowledge assets encompass documents, photographs, awards, CPARS reports, and press coverage.

3. Establish and communicate a proposal win bonus program (see Figure 1.5) to generate enthusiasm across the organization. Make working on proposals exciting and beneficial both financially and professionally within you company.

1.6 INSPIRING PASSION

Building systems for experimentation that help employees to discover their particular domains of interest and creating connections between workers are two important ways for organizations and managers to support passion [3]. To foster the establishment of connections among staff professionals, investment guidance company The Motley Fool employs a chief collaboration officer (CCO). In an empirical study of 381 small-scale businesses, the results suggested that passion rather than persistence contributes more to firm growth [4]. Sustained passion and proposal winning go hand in hand, and proposal winning propels business growth.

Figure 1.5 $10 bill. (Photograph © Dr. R. S. Frey.)

References

[1] Schultz, B., "7 Traits of High-Performing Acquisition Teams," *DAU,* July 1, 2019, https://www.dau.edu/library/defense-atl/blog/7-Traits-of-High-Performing-Acquisition-Teams-.

[2] Silverthorne, S., "Passion at Work Is a Good Thing—But Only If Bosses Know How to Manage It," *Harvard Business School Working Knowledge,* January 24, 2023, https://hbswk.hbs.edu/item/passion-at-work-is-a-good-thing-but-only-if-bosses-know-how-to-manage-it.

[3] Hagel III, J., and J. S. Brown, "How to Create a Workplace That Actually Inspires Passion," *Harvard Business Review,* July 23, 2020, https://hbr.org/2020/07/how-to-create-a-workplace-that-actually-inspires-passion.

[4] Iyortsuun, A. S., and C. Shakpande, "Passion, Persistence, and Firm Growth: Moderating Role of Environmental Uncertainty," *BRQ Business Research Quarterly,* 2022, p. 13.

2

PROPOSALING IN CONTEXT: WHERE DOES THE PROPOSAL FIT WITHIN THE ENTIRE MARKETING LIFE CYCLE?

2.1 INTRODUCTION

Most of the documents that we call proposals in the U.S. federal government marketspace come to be developed within a complex statutory and regulatory environment. Three-dimensional chess is child's play comparatively. The FAR is the primary regulation that all executive agencies, such as the Department of Health and Human Services (DHHS), Department of Veterans Affairs (VA), and Department of Energy (DOE), use to acquire goods and services with the funds that Congress appropriates based in part upon the President's detailed annual Budget Request. The FAR became effective on April 1, 1984, and is issued within applicable laws under the joint authorities of the Administrator of General Services, the Secretary of Defense, and the Administrator for the National Aeronautics and Space Administration (NASA). Through a series of circulars, guides, and memoranda, the Administrator of the Office of Federal Procurement Policy (OFPP) within the Office of Management and Budget (OMB) provides the direction for government-wide procurement procedures. The OFPP's overarching goal is "to promote economy, efficiency, and effectiveness in acquisition processes," as the *Federal Register,* the daily journal of the U.S. government, has affirmed.

Why do so many organizations of all sizes pursue federal contract dollars? The answer is simple—there is a tremendous amount of discretionary dollars that the government awards to contractors each year. To put a finer point on it, the Congressional Budget Office (CBO) reported that the federal government spent more than $1.7 trillion on these contracts in fiscal year (FY) 2022.

Even small businesses share in this amazing windfall. The Biden-Harris Administration awarded a record-breaking $154.2 billion in contracting to small businesses. The annual Small Business Federal Procurement Scorecard is an assessment tool to measure how well federal agencies reach their small business and socioeconomic prime contracting and subcontracting goals. These include goals for small businesses (SBs), women-owned small businesses (WOSBs), small disadvantaged businesses (SDBs), service-disabled veteran-owned small businesses (SDVOSBs), and small businesses located in Historically Underutilized Business Zones (HUBZones).

2.2 WRAPPING BUSINESS DEVELOPMENT, CAPTURE MANAGEMENT, AND PROPOSAL DEVELOPMENT IN A BLANKET

Lebanon Valley College near Hershey, Pennsylvania, was my undergraduate alma mater. Figure 2.1 is an image of a commemorative blanket from that small, private 4-year school. Now what does this blanket have to do with business development, capture management, and proposal development? Actually, a lot.

Think of the individual blue and white fringes on the edges of the blanket as strands of business intelligence that your company collects from listening to, and interacting with, the current or prospective federal government customer set. Answers to questions such as:

- What are the top three characteristics that make an outstanding industry partner for working with your organization?
- What is your vision for this contract 5 years from now?
- What are your primary concerns regarding the success of this contract going forward (e.g., funding levels, requirements de-

Figure 2.1 Blanket from Lebanon Valley College in Annville, Pennsylvania. (Photograph © Dr. R. S. Frey.)

scoping, platform migration, major facility move, or something else)?

- What specific value do you see in a contractor's project management office (PMO)?
- What types of technical innovations would you like to see introduced under this new contract?
- What levels and frequency of workload fluctuations have you experienced during the current contract? What do you anticipate moving forward?

The buildings represent the solutions that the capture management function has formulated based upon the business intelligence that has been collected and assessed. Competitive assessments also influence solutions. Think of solutions as the choices that your company makes from a technical, management, staffing, phase-in, and past performance standpoint. For example, will you use IBM Engineering Requirements Management DOORS as the requirements management tool, or, alternatively, Visure Requirements? That is a technical choice and, therefore, a technical solution. Will your company propose a deputy program manager as a key position or not? That is a management choice and, therefore, a management solution. Will your company offer to accelerate the phase-in schedule, or does your organization conclude that step introduces too much potential risk? That is a phase-in choice and, thus, a phase-in solution.

The blanket as a whole represents the overarching story that your company's proposal development staff professionals and subject matter experts weave from the threads of business development and the solutions of capture management. It is a tightly coupled story, replete with graphics and icons that have meaning in the customer's world.

To extrapolate to the winning proposal, the blanket provides financial "warmth" to your company for the period of performance of the contract.

2.3 EVERYONE IN YOUR COMPANY IS IN BUSINESS DEVELOPMENT

Fundamentally, there are six functions within your company that must communicate and work together well for sustained business success in the federal government marketspace. These encompass: (1) business development, (2) capture management, (3) proposal development, (4) KM, (5) infrastructure, and (6) operations. Notice that I specifically used the word "functions"—in many small organizations, for example, there are no dedicated staff who have the title of Business Developer or Capture Manager. That being said, everyone in your organization who is engaged in these functions should recognize that, no matter their title and reporting chain, they best serve your company and themselves professionally by recognizing that they are actually engaged in the arena of business development.

Knowing what you know as an organization—that is, the KM function, supports the first three functions named. For example, KM enables fact-based bid/no-bid and teaming decisions by providing insights into corporate contractual past performance and certifications, skillsets and competencies across your organization's staff professionals, quantified success stories from supporting specific government customers, and insights from Black Hat reviews regarding key competitors. Black Hat reviews focus on the strengths and weaknesses of your primary competitors, how they would likely position themselves to be awarded the contract, and how your company needs to shape its offer to best your competitors.

The KM function also supports proposal development with access to field-proven technical and management approaches, as well as documented plans, such as quality assurance (QA), safety and health (S&H), and diversity, inclusion, equity, and accessibility (DEI&A). These can serve as a meaningful starting point to support rapid proposal prototyping. Of note is that every proposal work product must be customized for the government customer and specific program being pursued.

Those staff who are engaged in recruiting, contract administration, invoicing, risk mitigation, QA, cost accounting, and legal review are also integral to business development success. For instance, recruiters work to locate and set up interviews with high-demand, low-density subject matter experts (SMEs) nationwide using job aggregator tools and talent sourcing platforms. They are often involved with orchestrating off-site open houses or technical interchange meetings (TIMs).

Operations staff, those folks who support the government customer every day, are in an ideal position to learn about upcoming transformational process initiatives or changes in operational cadence, as well as planned migration to a cloud-based platform such as Amazon Web Services (AWS) or Microsoft Azure. Importantly, this knowledge must be documented and collected in a systematic manner. Most definitely, the BD function should recognize the operations staff who identify opportunities and document those opportunities as part of the BD "funnel" of upcoming pursuits. The key is ongoing multilevel communication and prompt documentation.

As a company executive, one of your critical roles is to build, sustain, and recognize a winning culture across your organization. That winning culture results in part from clear communication from and active listening by leadership, ongoing teambuilding engagements, continuous learning and improvement, fostering a strong sense of job embeddedness, honoring diversity and inclusion, and providing tangible rewards for performance excellence and proposal wins. A publication by the Society for Human Resource Management (SHRM) introduced me to the concept of job embeddedness, first espoused in 2001 in an article entitled, "Why People Stay: Using Job

Embeddedness to Predict Voluntary Turnover," published in the peer-reviewed *Academy of Management Journal* [1].

Winning is a team sport (see Figure 2.2). Everyone must play for the entire team to win consistently.

2.4 LOOKING THROUGH THE LENS OF PMBOK

Proposals fit the Project Management Institute's (PMI) definition of a "project" to a tee—"a temporary endeavor undertaken to create a unique product, service, or result" [2]. Given this, let's take a look at business development, capture management, proposal development, and KM through the lens of the Project Management Body of Knowledge (PMBOK®) 6th Edition's framework of INPUTS, TOOLS & TECHNIQUES, and OUTPUTS (see Figure 2.3). We will start with business development.

Figure 2.2 Winning is a team sport. Three Colorado Rockies players pursue the one San Diego Padres' base runner. (Photograph © Dr. R. S. Frey.)

Figure 2.3 Macro lens. (Photograph © Dr. R. S. Frey.)

The INPUTS for business development in the U.S. federal government marketspace begin with your organization's strategic plan. That should serve as the primary guide for all of your company's new business pursuits. Other important INPUTS encompass Agency Recurring Procurement Forecasts (https://www.acquisition.gov/procurement-forecasts), System for Award Management (www.SAM.gov), Federal Agency Strategic Plans and Technology Roadmaps (e.g., 2020 NASA Technology Taxonomy and Department of the Air Force Posture Statement Fiscal Year 2022), and the Department of Defense (DoD) Office of Small Business Programs (https://business.defense.gov/), as well as insights from staff once employed by a specific federal customer organization, sponsored networking events, and intelligence learned from preferred teaming partners.

Among the TOOLS & TECHNIQUES associated with Business Development are an Opportunity Pipeline tracking tool (e.g., Capture2, Inc., Vantagepoint CRM Plus, and Bloomberg Government (BGOV)), developing a focused value proposition, call plan execution through one-on-one customer meetings, customized and succinct presentations, solutions-focused white papers, bid decision gate reviews, and bid and proposal (B&P) initiation forms.

The principal OUTPUT of the business development function is a fact-based business case that aligns with the company's strategic plan and that is presented to executive management. Additionally, OUTPUTS span fully populated bid decision gate reviews, carefully documented call plan reports, defined roles and responsibilities within the company for the specific federal opportunity, higher return on the B&P investment for the company, and inputs for the capture plan and proposal directive.

Of note is that your call plan reports should be structured in accordance with the expected major elements of the upcoming request for proposal (RFP). That is, business developers should invest the time to categorize their findings from each engagement with the customer into technical, management, staffing, and phase-in/phase-out (and other) "buckets." Even more beneficial to capture managers and proposal managers is for business developers to further demarcate their findings related to the technical bucket to a specific task or functional area from the previous or draft RFP or performance work statement (PWS).

The results of business development activities documented in call plan reports serve as the key INPUT for effective capture management.

2.5 USING THE LENS OF PMBOK TO FOCUS ON CAPTURE MANAGEMENT

Let's take a look at capture management through the lens of PMBOK 6th Edition's framework of INPUTS, TOOLS & TECHNIQUES, and OUTPUTS.

The INPUTS for capture management in the U.S. federal government marketspace span gate review decisions, call reports, and notes from government-conducted site tours and industry days.

Well-prepared call reports should include information gathered by business developers regarding the government customer's hopes, fears, biases, critical issues, and success factors. Note that "biases" are not negative, but rather refer to preferences that the customer has relating to technical and programmatic dimensions of the contract.

On a winning $400 million civilian agency proposal effort that I supported for nearly 2 years, I had the opportunity to be on the ground for both Industry Day and the 5-hour site tour in the Southwest United States. Among the key points that I learned directly was that the government was very much interested in engaging with industry as a partner, and that what-if scenario planning was an important activity. I saw the government organization's creed displayed on the walls of several conference rooms, and the message of safety and cost efficiency presented on impact posters. In addition, I heard a senior government manager build context in his own words for the organization within the agency-wide vision and mission, and also paid close attention to government staff voicing specific terms of art.

Other key INPUTS include results from data mining the government customer's technical library or portal, as well as Freedom of Information Act (FOIA) request results. Importantly, companies should FOIA recent monthly or quarterly Technical Progress Reports on the ongoing contract that is being pursued. There is a wealth of information regarding staffing levels, specific tasks, recurring deliverables, software tools, access for and involvement of foreign nationals, publications, and much more. Web-based research results can also serve to inform the capture effort.

Among the TOOLS & TECHNIQUES associated with capture management are a Plan of Actions and Milestones (POA&M); Strengths, Weaknesses, Opportunities, Threats (SWOT) analysis; International Organization for Standardization (ISO) 9001:2015-defined capture processes; and the capture plan. This plan should include capture team names and contact information, opportunity overview, description of the government customer's organization and environment, and competitive analysis. In addition, incorporate a profile of the notional winner, teaming strategies and rationale, technical and management solutions (i.e., choices), pricing strategies, key personnel selections and rationales, evidence of strengths, and relevant and referenceable past performance.

The principal OUTPUTS of the capture management function encompass an informed pursuit decision to present to your executive leadership, a 2 to 3-page Win Strategy white paper that captures the essence of your offer—aligned to the Evaluation Factors for Award, solution sets for all sections of the proposal, and clearly defined roles and responsibilities across the pursuit team. Recommend that you also include a RACI chart that codifies who is responsible and accountable for what, and who must be consulted and informed.

2.6 PROPOSAL DEVELOPMENT—PATHWAY TO SUCCESSFUL BLUE TEAM REVIEW

Let's use PMBOK® 6th Edition's framework of INPUTS, TOOLS & TECHNIQUES, and OUTPUTS to characterize the pathway to a successful Blue Team Review as part of proposal development (see Figure 2.4). From my perspective, the Blue Team constitutes a working, interactive meeting during which members of the Capture/Proposal Team present each one of the Blue Team work products to the Review Team. The discussion should include the rationales for specific solutions, that is, the choices that your organization is offering to the federal government customer across technical, management, phase-in/phase-out, staffing, and past performance.

The INPUTS for proposal development and the Blue Team process include the request for proposal (RFP) and attachments, including the past performance questionnaire (PPQ). INPUTS extend to your company's collective business intelligence (BI) about this program, associated technologies, stakeholders, governance, the government's future directions, and potential competitors. In addition, INPUTS encompass your firm's collective technical and programmatic successes on contracts of similar size, scope, and complexity; customer recognition/awards; Contractor Performance Assessment Reporting System (CPARS) reports; award fee scores; and your portfolio of past performance.

Among the TOOLS & TECHNIQUES are the following: (1) 3-column compliance matrix that includes a column for comments; (2) annotated outline; (3) validated strengths mapped to Evaluation Factors for Award in a table format; (4) SAR Model© Template for capturing and documenting success stories in the framework of

Figure 2.4 Large cobalt blue ceramic flower pot. (Photograph © Dr. R. S. Frey.)

situation-actions-results; (5) Proposal Readiness Work Products© that are used to codify understanding, approach, past performance validation, and value to the government; (6) stakeholder diagram (another PMBOK construct); (7) Win Strategy white paper; (8) calendar-style proposal response schedule; (9) gap analysis of BI; and (10) portal for knowledge transfer (KT) and communication.

OUTPUTS for presentation and review at the Blue Team include a populated compliance matrix, completed detailed annotated outline for each proposal volume, 2 to 3-page narrative-style Win Strategy white paper with an embedded table of "Evidence of Strengths" mapped to the evaluation factors for award, and fully populated SAR model success stories. In addition, OUTPUTS extend to a completed organizational chart for the program; fully populated Proposal

Readiness Work Products for the technical approach; completed stakeholder diagram to demonstrate understanding; calendar-style proposal response schedule; and past performance citations synopsized to indicate recency and relevancy (size, scope, and complexity), as well as documented BI gaps; and candidate questions to government.

Referencing another PMBOK construct—the Critical Path Method (CPM)—the Blue Team Review stands squarely on the critical path. Frankly, it is more important during the proposal development life cycle than most other color team reviews. Using the nomenclature of Lean principles, the Blue Team is an integral element of the Value Stream Map (VSM).

2.7 KM THROUGH THE LENS OF PMBOK

Now let's use PMBOK 6th Edition's framework of INPUTS, TOOLS & TECHNIQUES, and OUTPUTS to more closely examine KM.

The INPUTS for KM in a federal support services contractor span best-of-breed proposal sections, monthly operating reports, transcribed audio records from off-site managers' meetings, annual résumé updates, and semiannual project citation updates. In addition, source selection statements (SSS) and source selection decision documents (SSDD), after action reviews (AAR), and FOIA request results are important INPUTS. Also capture transcriptions of expert interviews and employee exit interviews, along with documents from government agency technical libraries, public-domain Federal Strategic Plans and Mission Statements, Government Accountability Office (GAO) reports, and agency-specific white papers, leadership biosketches, and testimony before Congress. Further, collect commendations received from your government customer, CPARS reports, industry awards and press coverage, and Black Hat review results on your competitors.

Among the TOOLS & TECHNIQUES are the following: (1) data warehouses, Content Services Platforms (CSP), and Document Management Systems (DMS) (e.g., Microsoft SharePoint 2021, Inforouter (Active Innovations), Privia, DocuShare (Xerox), Virtual Proposal Center (Intravation), Box, M-Files, and Google Drive); (2) knowledge codification to make knowledge easily available and transferrable; (3) communities of practice (CoP); (4) knowledge visualization (e.g.,

Google Chart, Tableau, Infogram); (5) storytelling to transfer knowledge in a contextual way that is also easy to understand; (6) knowledge maps; (7) knowledge retention tools (e.g., shadowing, wikis); (8) knowledge ontology—key process area to ensure a standard taxonomy (i.e., vocabulary, metadata) is used for classifying knowledge stored in the organization; (9) knowledge-sharing events (e.g., "pause and learn" moments); (10) knowledge acquisition—the result of successful KT; (11) data mining (e.g., RapidMiner Studio); (12) expertise locators; (13) visual assistants or chatbots; (14) knowledge audits; (15) Mitre's KM-Capability Maturity Model; and (16) KMAgile.

OUTPUTS from KM encompass enhanced evidence-based decision-making with relevant, data-led insights (e.g., bid/no-bid decisions); rapid proposal prototyping; leveraging existing expertise and experience; finding, sharing, and reusing intellectual capital; and avoiding redundant effort. Additional beneficial OUTPUTS include accelerated onboarding of new hires, creating new knowledge and insights, and increasing your company's adaptability to respond to

Figure 2.5 Handmade boat being completed at the Chesapeake Bay Maritime Museum in St. Michaels on Maryland's Eastern Shore. Note the wooden mosaic that tells a story. The wooden ribs represent all of the knowledge assets held within a structured framework. The boat will soon navigate across VUCA waters. (Photograph © Dr. R. S. Frey.)

volatility, uncertainty, complexity, and ambiguity—VUCA[1]—in the federal marketspace (see Figure 2.5).

References

[1] Mitchell, T. R., et al., "Why People Stay: Using Job Embeddedness to Predict Voluntary Turnover," *Academy of Management Journal,* 2001, https://journals.aom.org/doi/10.5465/3069391.

[2] Project Management Institute, *A Guide to the Project Management Body of Knowledge (PMBOK®), Seventh Edition,* Newtown Square, PA: Project Management Institute, p. 245.

1. Coined by anthropologist, historian, and futurologist Jamais Cascio, the term brittle, anxious, nonlinear, and incomprehensible (BANI) has emerged following the COVID-19 pandemic.

3

THE IMPORTANCE OF HUMAN DYNAMICS

3.1 METAPHORS MATTER

Successful proposaling results from focused passion and insightful creativity. Yet the overwhelming majority of organizations with which I have worked directly—from *Fortune* 50 multinationals with staff numbering in the tens of thousands to 45-person small businesses headed by a single entrepreneur—perennially talk in terms of "proposal engine," "prop shop," and "cranking out proposals," as if proposal volumes built with interchangeable parts proceed along an assembly line. Touched by proposal technicians along the way toward a final electronic "white glove" inspection before being uploaded to a government-wide acquisition contract (GWAC) acquisition portal, such as the General Services Administration's (GSA) e-tool called eBuy, or, much less prevalent in 2023, printed, loaded into 3-ring binders, boxed, and hand-delivered to the appointed delivery office at a brick-and-mortar government office building or military facility by the date and time stipulated in the RFP or request for quotation (RFQ). This very scenario occurred with a Department of Veterans Affairs (VA) proposal that had to be delivered in person and in hardcopy notebooks to Golden, Colorado, in October 2021.

Making a positive difference is what we all want to do within our businesses and other organizations. However, sometimes words, which, in turn, drive human perceptions, get in the way of optimizing the positive difference that individual proposal professionals, and colocated and virtual proposal teams, can make.

After working across 104 different organizations during the past 16 years in my own consultancy, too often I hear executive managers talk in terms of aspiring to have their proposal staff function as a "well-oiled machine" (see Figure 3.1). This metaphor conjures up ideas of interchangeable and replaceable cogs. Mechanistic models associated with early twentieth-century scientific management thinking—such as Frederick Taylor's time and motion studies in factory settings—tend to induce people to behave in predictable ways. These

Figure 3.1 Hot rod engine on display on the Las Vegas Strip. (Photograph © Dr. R. S. Frey.)

models are best suited for organizational components that experience stable, relatively unchanging environments.

As you certainly know, that is not the way that the federal or commercial proposal worlds work today. Changing the metaphors by which proposal teams are perceived, as well as the lens through which they view themselves, will be valuable to federal and commercial contractors.

In order to respond to these procurements, companies must migrate beyond the machine models for their proposal staff and encourage innovation, speed, and dexterity.

We should describe proposal professionals as value-added, knowledge-rich service providers. If we change the way that we talk about proposal teams, employees could change their behavior in a way that will drive up probability of win (P_{win}), increase productivity, generate and implement innovative ideas, and enhance staff retention and institutional knowledge. New metaphors could also raise the profile and value of proposal professionals within a given organization. Executive leadership and other stakeholders will certainly reap the benefits of igniting and promoting this profound shift.

In moving forward to face the challenges of significant change within the federal and commercial procurement environments—where innovation, speed, knowledge-sharing, and collaboration are critical success factors—it is time to change the optics of our perception. "Engines" are fundamentally not adaptive, and "commodities" are inherently not innovative.

However, proposal team social systems will be able to shift and evolve to thrive in the evolving federal and commercial contracting environments. Human-centric metaphors such as social systems promote flexibility. People can initiate change and adapt and adjust rapidly to evolving conditions like those we see across the public and private-sector space. As evidenced by the COVID-19 pandemic, we live in a volatile, uncertain, complex, and ambiguous (VUCA) world; the term VUCA was introduced by the U.S. Army War College located in Carlisle, Pennsylvania. The metaphor of proposal team as an adaptive and dynamic social system encourages innovative behavior.

This, in turn, contributes to professional and personal growth. It also encourages "organizational ambidexterity" (a term used by Stanford University's Charles A. O'Reilly III and Harvard Business

School's Michael L. Tushman) and corporate-level success, as measured by stakeholder return on investment (ROI) and/or the balanced scorecard (BSC). Developed by Dr. Robert S. Kaplan and Dr. David P. Norton, the BSC strategic planning and management framework considers financial performance, customer satisfaction, efficiency of internal business processes, and organizational capacity, which encompasses innovation.

So, with these thoughts in mind, what if executive leadership used new metaphors to identify and characterize their proposal groups, such as proposal knowledge teams (PKTs), proposal innovation centers (PICs), or proposal knowledge integrators (PKI). These forward-leaning organizational titles—and the mental pictures or symbols that they put forward—would serve to elevate the value of the professionals who comprise them—as they certainly deserve to be. These new titles will help to drive positive perception and, importantly, the business, professional, and interpersonal actions that follow directly on the heels of that new perception.

Indeed, metaphors still matter.

3.2 THE 71% EFFECT

In Figure 3.2, a flock of Canadian geese are shown flying in a V formation over Blackwater National Wildlife Refuge on Maryland's Eastern Shore. Now what do these birds have to do with business development, capture management, proposal development, operations, and infrastructure success for a federal support services contractor in 2023? Actually, quite a lot.

In 1970, Peter Lissaman and Carl Shollenberger published an article in the peer-reviewed journal, *Science*. This was the first article to detail the precise aerodynamic interactions that were likely taking place within a flock that could produce an energetic benefit. The reduction in power requirements equate to an increase in flight range for formation flight versus solo flight [1]. Using a computer model developed by aviation engineers, ornithologists found that birds flying in Vs could realize an energy savings of 71%, compared to birds flying alone [2]. Importantly, the V formation may also improve the communication between individual birds [3].

Figure 3.2 Flight of geese over the Blackwater National Wildlife Refuge on Maryland's Eastern Shore. (Photograph © Dr. R. S. Frey.)

Let's transfer these insights into the business world. There is significant competitive advantage to be gained when everyone in the organization moves forward together toward shared corporate goals. How does this happen? One way that I have seen work extraordinarily well is by conducting semiannual strategic reviews. Over the course of my nine-year tenure with a small firm in Northern Virginia, these strategic reviews involved the active engagement of corporate executives, business development, division vice presidents, technical managers (operations), proposal development, infrastructure members, and select administrative staff. Reviews were convened at an off-site location to enable complete focus and participation.

These retreats were useful for direct information sharing on preplanning, as well as updates on active business development targets and current proposal activities. I found the updates on project-level status and achievements across the company to be proposal gold. Well beyond focusing on budget goals only, strategies and tactical objectives for the upcoming 6 months were presented, discussed, and documented. Everyone came away energized and informed. Communication pathways were established across organizational lines, and friendships were built and sustained. We knew we were in it together—and the flight was genuinely amazing. Institute the V in your company. It works.

3.3 STOP DROWNING IN MEETINGS

I notice that particularly when working with large corporations, the number, frequency, and duration of preproposal and proposal meetings are literally overwhelming (see Figure 3.3). Now, I certainly understand the need for multidimensional communication to ensure awareness, engagement, schedule adherence, and tangible progress on action items and deliverables. Ongoing communication is also critical to encourage teamwork and promote synergy, particularly in virtual or hybrid working environments across large or geographically dispersed teams. However, my direct observation is that some meetings are scheduled in an ad hoc manner, which can lead to scheduling conflicts. Suddenly, a meeting invitation appears through Microsoft Teams, for example, with little advance notice (20 minutes prior is not uncommon). Planned work gets interrupted. In addition, many meetings are not accompanied with a meaningful agenda sent in advance that lists specific outcomes, as well as roles and responsibilities of the participants. Frequently, participants—most notably executive leadership—join a meeting late and have to be back-briefed. This wastes everyone's time. Then, in the chat window, there is the inevitable post: "I need to drop to join another meeting."

There is software such as Microsoft Teams, Asana, Monday.com, and Google Tasks to track action items that surface during the course of a meeting. However, too often I see that during a focused action item review, members of the capture and proposal team cannot recall the exact context and scope of a given action item. This may result in

Figure 3.3 Aerial view of waves along Sunset Cliffs on Coronado Island near San Diego, California. (Photograph © Dr. R. S. Frey.)

a spin-off meeting to determine exactly what was meant. Yet another meeting.

Steven G. Rogelberg, Chancellor's Professor and professor of management at the University of North Carolina in Charlotte, authored a book called *The Surprising Science of Meetings: How You Can Lead Your Team to Peak Performance* [4]. In this work, Professor Rogelberg included a meeting quality self-assessment designed to examine an organization's ROI from its meetings. I would strongly recommend that organizations of all sizes conduct and then take action on the results of this internal organizational reflection [5].

I have seen the technique called "timeboxing" applied very effectively in meetings. A term first coined by Scott Schultz and James Martin, the *Agile Alliance* defined a timebox as [6]: "a previously agreed period of time during which a person or a team works steadily towards completion of some goal." The meeting leader or moderator should inform the group about the time increment that they have for a given discussion topic and then end the discussion when that time period has elapsed. When action items emerge, ensure that someone is assigned to capture detailed, intelligible notes with context built in.

Time is every business's most valuable resource. Optimize the value of each meeting. Ensure that meetings advance the process of delivering a compliant, credible, compelling, and cost-competitive proposal.

3.4 SEARCHING FOR EFFECTIVE COMMUNICATION

In 2023, I was working on a $2 billion civilian agency proposal with a major business consulting services firm. The corporation had multiple large and small business teammates, which is typical for complex, enterprise-wide opportunities. To firewall competition-sensitive information from teammates and select consultants, my customer leveraged both Privia and Microsoft Teams permissions-based platforms to house proposal-related information. I observed that there was some uncertainty as to which platform the proposal documents of record were stored.

On any given day, a member of the capture and proposal team would receive emails, texts, and instant messages (IMs). Add to that, there were announcements, wikis, chat notices, and posts accessible through Microsoft Teams. Additional communication could occur through the "Activities" icon, and information could also be shared through the "General" channel. Those are 9 unique communication pathways. Priorities are overtaken by other higher priorities, and ad hoc meetings emerge at any time.

Several other challenges have emerged. The first is precisely where on Microsoft Teams critical folders and files are stored. The location of the "wall of truth" that represents a compendium of customer insights and company approaches and metrics proved elusive for me even after 3.5 months of preproposal support to this company. From a knowledge management perspective, the link to that information should be pushed out by the proposal coördinator to all members of the capture and proposal team on a regular basis. Why? Because the wall of truth is dynamic; it is changing constantly. That same outreach action should be done with staff working on the total compensation plan (TCP), for example. People are added to or leave the TCP subgroup. Providing the link to the working folders related to TCP each week would have helped to ensure that all contributors knew where to go on Microsoft Teams to find the latest relevant files.

A second challenge is related to internal company practices and nomenclature. Every organization has unique terms and acronyms for its review cycles (e.g., Yellow Team, Silver Team, Synergy Review, Step Review, and Executive Solution Review (ESR)). This issue relates to teammates and outside consultants understanding the structure, participants, and outcomes of those reviews. There should be a READ ME or ROADMAP folder on Microsoft Teams in which that information is collected and company-specific acronyms are defined.

Food for thought.

3.5 ORGANIZATIONAL CULTURE AND PERFORMANCE SUCCESS

A large body of research suggested that "culture is the driver of business performance" [7]. In addition, these researchers assert that culture is the "driver of competitive advantage." Most scholars agree that the concept of organizational culture includes both tangible and intangible elements. One working definition [8] is that organizational culture constitutes the "set of basic assumptions or beliefs and values which determine the behavior, as well as the mode of operation and activities of an organization."

I have worked full-time in a highly successful small business support services contractor for the U.S. federal government. The organizational culture was shaped by a strong and charismatic president, a quality-driven and customer-focused chief operating officer, and a risk-taking, barrier-surmounting vice president for business development. Everyone was encouraged to be an entrepreneur. In fact, the company presented a striking Entrepreneur of the Year award at the holiday party each December. Additionally, this organization built comradery across the entire staff through community service initiatives that the executive team spearheaded and in which they rolled up their sleeves to paint or do yardwork. The organizational culture functioned as the social glue [9] that bound the employees and managers together and made them feel they were an integral part of the positive organizational experience. I can honestly say that I have never felt more a part of an organization than that particular one.

"In the age of drastic changes[,] organizations need sustainable competitive advantage to cope with changes and to succeed" [10]. I would submit that a strong and positive organizational culture helps to blunt the impact of environmental turbulence characterized as VUCA. More recently, the framework BANI (brittle, anxious, nonlinear, and incomprehensible) has been applied to the challenges presented by a rapidly changing business environment (see Figure 3.4). This term originated with Jamais Cascio, an American anthropologist, futurist, and author.

Organizational performance can be categorized into two categories: financial and nonfinancial. Earnings before interest, taxes, depreciation, and amortization (EBITDA) is a widely used measure of corporate profitability. Nonfinancial measures encompass employee job satisfaction and voluntary employee turnover totals.

As leaders and managers, I recommend that you invest your time and energy into building and nurturing an inclusive business culture that welcomes diversity and innovative thinking to catalyze success. I resonate with the notion of culture as a tool kit, which was advanced by Stanford University's Dr. Ann Swidler [11]. The cultural tool kit contains "symbols, stories, rituals, and world-views, which people may use in varying configurations to solve different kinds of prob-

Figure 3.4 Thrill-seekers travel upside down on a roller coaster at California's Great America amusement park in Santa Clara. (Photograph © Dr. R. S. Frey.)

lems." In this light, "culture" has similarities to the Agile mindset, with its "stories" and "ceremonies" (i.e., "rituals").

Keep your company's collective cultural tool kit brimming with possibilities.

Selected Bibliography for Building Effective Teams

The following sources are provided to help you as leaders and managers to build effective teams in your own organizations:

Coyle, D., *The Culture Code: The Secrets of Highly Successful Groups,* New York: Bantam Books, 2018.

Gordon, J., *The Power of a Positive Team: Proven Principles and Practices That Make Teams Great,* New York: Wiley, 2018.

Lencioni, P., *The Five Dysfunctions of a Team: A Leadership Fable,* New York: Jossey-Bass, 2002.

Pritchard, K., and J. Eliot, *Help the Helper: Building a Culture of Extreme Teamwork,* New York: Portfolio/Penguin, 2012.

Simon, P., *Reimagining Collaboration: Slack, Microsoft Teams, Zoom, and the Post-COVID World of Work,* New York: Motion Publishing, 2021.

Tannenbaum, S., and E. Salas, *Teams That Work: The Seven Drivers of Team Effectiveness,* New York: Oxford University Press, 2021.

Wheelan, S., M. Åkerlund, and C. Jacobsson, *Creating Effective Teams: A Guide for Members and Leaders,* Sixth Edition, New York: Sage, 2021.

References

[1] Portugal, S., "Lissaman, Shollenberger and Formation Flight in Birds," *Journal of Experimental Biology,* Vol. 219, No. 18, 2016, pp. 2778–2780.

[2] Wilson, H., "Vee Formations and Bird Migration," *Maine Birds,* January 2016.

[3] Kölzsch, A., et al., "Goose Parents Lead Migration V," *Journal of Avian Biology,* January 2020.

[4] Rogelberg, S. G., *The Surprising Science of Meetings: How You Can Lead Your Team to Peak Performance,* New York: Oxford University Press, 2019.

[5] https://www.stevenrogelberg.com/meeting-quality-assessment-tool.

[6] https://www.agilealliance.org/glossary/timebox/.

[7] Gallagher, S., C. Brown, and L. Brown, "A Strong Market Culture Drives Organizational Performance and Success," *Employment Relations Today,* 2008, pp. 25, 27.

[8] Xanthopoulou, P., A. Sahinidis, and Z. Bakaki, "The Impact of Strong Cultures on Organizational Performance in Public Organizations: The Case of the Greek Public Administration," *Social Sciences,* Vol. 11, No. 10, 2022, p. 5.

[9] Akpa, V. O., O. U. Asikhia, and N. E. Nneji, "Organizational Culture and Organizational Performance: A Review of Literature," *International Journal of Advances in Engineering and Management (IJAEM),* Vol. 3, No. 1, 2021, p. 365.

[10] Aghazadeh, H., "Strategic Marketing Management: Achieving Superior Business Performance Through Intelligent Marketing Strategy," *Procedia - Social and Behavioral Sciences,* Vol. 207, 2015, p. 125.

[11] Swidler, A., "Culture in Action: Symbols and Strategies," *American Sociological Review,* Vol. 51, 1986, p. 273.

4

PROPOSAL DEVELOPMENT AND KNOWLEDGE MANAGEMENT

4.1 COLLABORATION TOOLS

Today, there is a boatload (see Figure 4.1) of permissions-based software applications, platforms, and services that companies use to support proposal development: Active Innovations' InfoRouter, Microsoft SharePoint, Box for Business, Xait's Privia, and Intravation's Virtual Proposal Center (VPC), to name several that I have used personally, along with Google Drive and Xerox DocuShare. While these tools can certainly be of benefit from the perspective of collaboration, document management, and proposal security, implementing and administering them for effective and efficient proposal use are too often lacking. These tools must be configured to be focused on the end user, not the needs of the prime contractor's information technology (IT) department. Importantly, using one of these tools is not the equivalent of genuine KM—a critical success factor in winning proposal development. Portals become data dumping grounds that frequently render them unusable and definitely not the #1 place to go for solution architects, technical and programmatic content providers, or Color Team reviewers on a given proposal opportunity.

Figure 4.1 Boat moored in the harbor at Portland, Maine. (Photograph © Dr. R. S. Frey.)

Having developed the curriculum for and taught organizational knowledge management (OKM) at the graduate level at Kent State University in Ohio, I came to recognize that organizational maturity, insofar as KM is concerned, involves moving beyond the concept of "knowledge is power" to "knowledge sharing is power." KM is all about getting the right information to the right people on the proposal team at the right time. At the most fundamental level—that of a mathematical equation—KM is a dynamic combination of structured processes and automated tools multiplied by executive-level leadership and vision, and, in turn, leveraged exponentially by passionate people and communities of practice (CoPs) or user groups.

Too frequently, I observe and experience organizations allowing these collaborative tools to impede the proposal development process.

It can take literally days for everyone on the proposal team to be given appropriate permissions. File naming and structure need to be clear and intuitive. The proposal manager or proposal coordinator should consistently push out guidance to the proposal team in the use of the portal or platform. Individual content providers need to be informed as to which of the potentially hundreds of documents they should focus their time and attention.

Leverage collaborative tools to the fullest, while keeping in mind that they are there to increase productivity and creativity. Avoid having the tools stand in the way.

4.2 SPINNING PROPOSAL GOLD

Most federal support services contractors with whom I have interacted professionally conduct monthly or quarterly reviews of the portfolio of their programs and projects. Generally, the focus tends to be on financials, staffing, key schedule milestones, "blockers" in Agile parlance, and open issues. Recommend that your organization consider adding several elements to these regular reviews that will help capture and codify important information for future use in proposal development, for example, (1) success stories using the situation, action, results (SAR) framework, (2) customer recognition in the form of emails or texts, and (3) staff publications or presentations. The SAR framework for developing proof points and validation, which I developed and copyrighted, looks like this:

- *Situation:* Characterize the program-level, project-level, or task-level situation or challenge that your government or private-sector customer faced. Do this characterization in one brief, clear, and compelling sentence.
- *Action:* Succinctly describe the particular actions that your organization took to solve the customer's challenges or meet their request. Be specific insofar as tools, techniques, processes, subject matter expertise, staffing mix, and/or innovations applied.
- *Results:* Describe the results that your organization achieved for your customer—both quantitatively and qualitatively. Describe how your company's actions resulted in improvements for the customer in the terms of enhanced performance quality,

schedule adherence or compression, cost control or avoidance, and/or risk mitigation.

Another outstanding but often overlooked source of proposal gold is the semiannual or annual off-site senior leadership meetings. These are the ones that involve an organization's program-level and project-level managers, as well as executive staff. Whether in-person or virtual, be sure to include representation from your proposal development staff in these meetings. Their presence and participation contribute to their learning more about your organization's customers and operations. This, in turn, helps them become even more valuable assets for tailoring past performance citations, as one example. In addition, consider video or audio recording the proceedings and then having them transcribed so that the information becomes easily stored and searchable in your proposal knowledge base. Frequently, there are one-of-a-kind statements made at this type of meeting that no one can actually recall, but know it was stellar. Recording preserves those strands of information and insight that can be spun into proposal gold (see Figure 4.2).

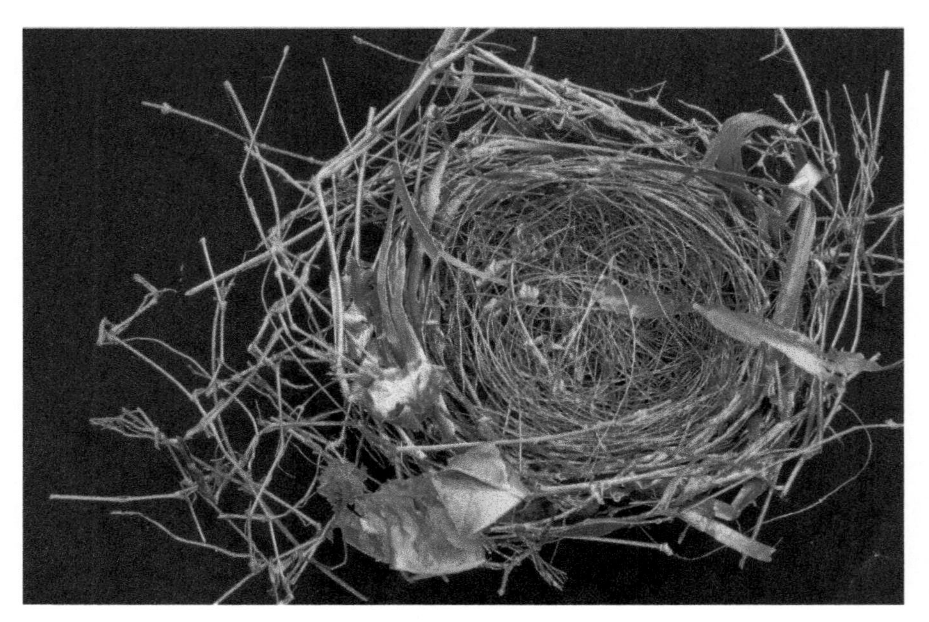

Figure 4.2 Actual bird's nest sprayed with gold paint. (Photograph © Dr. R. S. Frey.)

4.3 AGILE PROJECT MANAGEMENT AND EFFECTIVE KNOWLEDGE SHARING IN ORGANIZATIONS

What is the impact and value of Agile Project Management (APM) on knowledge sharing in an organization? That was my research question for the course, "Agile Project Management," through Harvard University Extension School in 2020. With its roots in the Agile Manifesto, APM has been introduced beyond software development to manage changing priorities, accelerate project delivery, and enhance delivery predictability, as well as better manage teams, reduce project risks, and drive down project costs [1].

A comprehensive review of qualitative literature from multiple geographic locations, including Thailand, Sweden, Germany, Ireland, South Africa, Brazil, and the United States, suggests that Agile Project Management, Agile Software Development, and Scrum practices foster knowledge sharing, particularly in terms of tacit knowledge transfer. Tacit knowledge and explicit knowledge were first described in 1958 by Michael Polanyi, a Hungarian-British philosopher of science [2]. In KM, tacit knowledge refers to knowledge that is known only by an individual and is difficult to communicate to the rest of an organization (see Figure 4.3). Knowledge that can be communicated more easily or is documented in a Word file or podcast, for example, is called explicit knowledge.

According to the American Productivity & Quality Center (APQC), the world's foremost authority in benchmarking, process and performance improvement, and KM, Agile benefits KM by integrating user needs, decreasing cycle time, and delivering tangible results [3]. In Agile environments, cross-functional teams are used to facilitate better knowledge sharing. The knowledge transfer chains are shortened by direct communication and collaboration [4]. Scrum activities, such as daily stand-ups, iteration planning, release planning, iteration reviews, and retrospectives support the growth of individual knowledge because individuals learn from experiences and knowledge shared during meetings. Moreover, these practices allow sharing and discussion among team members [5, 6]. The Agile cultural infrastructure includes values such as cooperation and knowledge sharing.

Visualization is a communication technique used to manage knowledge in a visible and accessible form. This strategy helps Agile

Figure 4.3 Sculpture in San Diego, California, called "Pacific Soul" crafted by artist Jaume Plensa. (Photograph © Dr. R. S. Frey.)

teams to support tacit knowledge sharing among team members. For example, the Scrum board is used to show progress based on the story cards; the whiteboard for information such as feedback, feelings, and action points in the retrospective meetings [7], the burndown chart for showing team progress, achievements, and performance, and the software code on display screens for showing working code in pair programming. The most commonly used visualization strategy in Agile practices is the Scrum board that contains all user stories and presents progress through the work status of team members [7, p. 203].

The extent to which employees understand and apply Agile knowledge-sharing methodologies at work was found to be correlated with the ability of the firm to experience enhanced capabilities of communication and coordination, as well as its ability to use knowledge for the development of new products or services [8], and the creation of new knowledge.

4.4 LIGHTING THE WAY—GETTING THE MOST OUT OF WHAT YOU KNOW

Just as I have suggested that everyone within an organization should perceive themselves as and, in turn, be seen as involved in business development in some manner, the exact same recommendation pertains to KM (see Figure 4.4). It is too easy for managers and members of a company to conclude that KM processes are the sole domain of dedicated knowledge-focused staff or KM champions. Everyone with-

Figure 4.4 Sparks shoot from the stone wheel of a shop grinder. (Photograph © Dr. R. S. Frey.)

in the business should be responsible for, and appropriately recognized for, sharing and supporting the capture and transfer of knowledge. This encompasses documenting "know what," "know why," and "know who." Each person has important knowledge to share and can benefit from the knowledge of others, as discussed in [9].

Insights gained from the APQC Open Standards Benchmarking database [10] show that KM programs report predominantly to the chief information officer (CIO) or chief human resources manager. A much smaller number of KM programs report to IT or information strategy organizational elements. I see that as a good thing because KM is far more than being only about software and technology. However, there must be a strong collegial working relationship between internal KM-focused staff professionals and IT subject matter experts (SMEs).

My own direct experience inside of a federal support services contractor showed me that anchoring the KM function within the business development group was a clear pathway to sustained success. Importantly, the KM function received direct executive-level advocacy and prioritization, financial support, and hands-on engagement. Knowledge sharing was also built into position descriptions and assessed during performance reviews. Clearly, this small business contractor stood high on any knowledge management maturity models (KMMM). There are actually numerous KMMMs, some of which are based on the Capability Maturity Model (CMM) developed by the Software Engineering Institute (SEI) at Carnegie Mellon University (CMU). For a very recent discussion of KM maturity frameworks, see [11].

One particularly intriguing KMMM that follows the Capability Maturity Model assesses 8 structural fields of KM: (1) leadership, support; (2) collaboration, culture; (3) people, competencies; (4) environment, partnerships; (5) strategy, knowledge objectives; (6) processes, roles, organization; (7) technology, infrastructure; and (8) knowledge structures, knowledge forms [12].

When I led the launch and establishment of our company's KM initiative, it would have been helpful to have had the newly developed and revised standard: ISO 30401:2018/Amd 1:2022 [13]. Published in April 2022, this document "sets requirements and provides guidelines for establishing, implementing, maintaining, reviewing and improving an effective management system for knowledge management in organizations" [13].

4.5 ARTIFICIAL INTELLIGENCE AND KM IN PROPOSAL LAND

Both KM and artificial intelligence (AI) can contribute to "rapid proposal prototyping"[1] and enhanced business decision support. "At their core, KM and AI are about knowledge ... KM enables an understanding of knowledge, while AI provides the capabilities to expand, use, and create knowledge" [15]. However, as Sangzoni et al. [16, p. 4] asserted, AI will not eventually subsume KM in such a way that "all tasks could eventually be algorithmically performed through codified explicit and tacit knowledge forms." This is precisely because "[s]ome forms of tacit knowledge are feasible to codify, and others aren't" [16, p. 6]. There is also the dimension of knowledge context, for which humans are highly adept [17, p. 93]. To be sure, AI can indeed adequately support KM "when dealing with explicit knowledge ... as complementary prostheses of human capabilities" [16, p. 15–16]. Helpfully, Jarrahi et al. [17, p. 91] have spoken of "collaborative intelligence, in which AI and humans enhance each other's complementary strengths."

Jarrahi et al. [17, p. 89] identified potential AI applications across different KM processes. For example, in knowledge creation, AI can be leveraged to recognize previously unknown patterns and to discover new relationships within large datasets. In knowledge storing and retrieving, AI can support "[h]arvesting, classifying, organizing, storing, and retrieving explicit knowledge." In addition, AI technologies are capable of "generating know-who (i.e., sources of expertise) within and across organizational boundaries..." [17, p 93].

As knowledge workers, proposal professionals will most certainly need to develop AI literacies going forward. These literacies encompass "algorithmic competencies ... and new analytical, data-centered skills that help workers interpret AI-based decisions" [17, p. 95]. AI writers such as ChatGPT or Google Bard are tools that rely on "artificial intelligence to automatically generate written content" [18]. Most definitely, these AI writers will be leveraged to compose

1. This term was first introduced by Dr. R. S. Frey in [14].

proposal narrative across technical, management, staffing, and phase-in/phase-out sections in the near future. From a career development perspective, proposal professionals will necessarily need to avoid cognitive complacency, while building a symbiotic relationship with intelligent systems [17, p. 97]. "[H]uman judgment is necessary in order to interpret and filter the AI-generated information" [19, p. 408]. This is because "[h]uman reasoning is able to make use of a wide context of human experiences, backgrounds, and skills and bring this to bear in solving business problems; in contrast, AI systems typically have a very narrow focus" [19, p. 411]. It will be important for company leadership and their proposal staff to recognize and value the complementarity that human intelligence and AI share.

Among the advantages of AI for KM are the following 8 benefits: (1) providing easier knowledge delivery; (2) removing linguistic barriers; (3) enhancing intellectual capital; (4) offering a real-time KM system; (5) enhancing knowledge sharing, usage, and capture; (6) enhancing flexibility of knowledge representation; (7) providing knowledge based on personal preferences; and (8) offering easy problem-solving [20, p. 4]. "Organizations can use intelligent agents from various AI-related technologies, such as genetic algorithms, intelligent agents, and neural networks, to perform tasks such as ... semantic analysis of texts, text mining, and pattern matching" [20, p. 5]. "Semantic analysis is the process of drawing meaning from text. It allows computers to understand and interpret sentences, paragraphs, or whole documents, by analyzing their grammatical structure, and identifying relationships between individual words in a particular context" [21].

Document AI leverages machine learning (ML), natural language processing (NLP), and intelligent character recognition (ICR) to extract unstructured information contained in digital and printed documents as well as "non-numerical and multifaceted" [22] data "in the form of text, audio, or images" [19, p. 405]. One example is Google's Document AI, which leverages computer vision, NLP, ML, and deep learning. Another example is Amazon's Textract [23]. Both of these tools can be leveraged to assemble and sustain proposal knowledge bases.

Blending the power of AI with the contextualized insight of human proposal talent will allow companies to increase the throughput of their proposal development organizations. This accelerated throughput must necessarily be driven by informed bid/no-bid decision-making.

References

[1] Ciric, D., et al., "Agile Project Management Beyond Software Development: Challenges and Enablers," *9th International Scientific and Expert Conference,* Novi Sad, Serbia, October 10–12, 2018, pp. 245, 249, https://www.researchgate.net/publication/351811492_Agile_Project_Management_beyond_Software_Development_Challenges_and_Enablers.

[2] Sidenvall, A., "Knowledge Sharing in and Between Agile Software Development Teams Using Knowledge Practices: An Interpretive Case Study at a Medium-Sized Medical IT Company," Unpublished Master's Thesis, Linköping University, 2017, p. 20, http://liu.diva-portal.org/smash/get/diva2:1115507/FULLTEXT01.pdf.

[3] APQC, "Applying Agile in Knowledge Management," APQC, 2019, p. 3, https://www.apqc.org/resource-library/resource-listing/applying-agile-knowledge-management.

[4] Melnik, G., and F. Mauer, "Direct Verbal Communication as a Catalyst of Agile Knowledge Sharing," *Agile Development Conference,* Salt Lake City, UT, June 22–26, 2004, p. 23.

[5] Neto, F. S., "Impact of Agile Practices on Organization Learning: A Model for Knowledge Creating and Sharing in Agile Teams," Unpublished Master's Thesis, FUMEC University, 2016, p. 46.

[6] Cognizant, "Knowledge Management in Agile Projects," White Paper, 2011, p. 3, https://www.cognizant.com/InsightsWhitepapers/Knowledge-Management-in-Agile-Projects.pdf.

[7] Andriyani, Y., R. Hoda, and R. Amor, "Understanding Knowledge Management in Agile Software Development Practice," in G. Li, et al. (eds.), *Knowledge Science, Engineering and Management (KSEM 2017),* Lecture Notes in Computer Science, Vol. 10412, 2017.

[8] Palminteri, M. R., and C. Wilcox, "Knowledge Sharing in an Agile Organization: As Enhancer of Dynamic Capabilities and Enabler of Innovation," Unpublished Master's Thesis, Blekinge Institute of Technology, Sweden, 2017, p. 43, http://www.diva-portal.se/smash/get/diva2:1119826/FULLTEXT02.pdf.

[9] *Guide for Building and Strengthening Organizational Knowledge Management Capacity in Organizations Working in Global Health,* USAID 2021, https://www.kmtraining.org/wp-content/uploads/2022/06/guide-building-strengthening-km-capacity-2021.pdf.

[10] Trees, L., "What Are the Best Knowledge Management Reporting Relationships?" *The APQC Blog,* August 13, 2018, https://www.apqc.org/blog/what-are-best-knowledge-management-reporting-relationships.

[11] Bougoulia, E., and M. Glykas, "Knowledge Management Maturity Assessment Frameworks: A Proposed Holistic Approach," *Knowledge and Process Management: The Journal of Corporate Transformation,* 2022, pp. 1–32, https://onlinelibrary.wiley.com/doi/epdf/10.1002/kpm.1731.

[12] Ehms, K., and M. Langen, "Holistic Development of Knowledge Management with KMMM®," Siemens, 2002, http://www.kmmm.org/objects/kmmm_article_siemens_2002.pdf.

[13] ISO 30401:2018/Amd 1:2022, Knowledge Management Systems—Requirements—Amendment 1, April 2022.

[14] Frey, R. S., "Small business knowledge management success story—This stuff really works!," *Knowledge and Process Management,* Wiley, 2002, Vol. 9, No. 3, p. 172, https://doi.org/10.1002/kpm.147.

[15] Smuts, H., and C. Borgstein, "Artificial intelligence (AI) and knowledge management (KM): Two sides of the same coin?," *Knowledge Management South Africa (KMSA),* November 14, 2021, p. 2, https://realkm.com/2021/11/14/artificial-intelligence-ai-and-knowledge-management-km-two-sides-of-the-same-coin/.

[16] Sangzoni, L., G. Guzman, and P. Busch, "Artificial Intelligence and Knowledge Management: Questioning the Tacit Dimension," *Prometheus–Critical studies in Innovation,* 2017, DOI:10.1080/08109028.2017.1364547.

[17] Jarrahi, M. H., et al., "Artificial Intelligence and Knowledge Management: A Partnership Between Human and AI," *Business Horizons,* Vol. 66, 2023, https://www.sciencedirect.com/science/article/pii/S0007681322000222.

[18] Henderson, R., "Chat GPT, the AI Writer - A Tool or a Weapon?," *Capital Chronicles,* April 15, 2023, https://www.capitalchronicles.ca/post/chat-gpt-the-ai-writer-a-tool-or-a-weapon.

[19] Paschen, J., M. Wilson, and J. J. Ferreira, "Collaborative Intelligence: How Human and Artificial Intelligence Create Value Along the B2B Sales Funnel," *Business Horizons,* Vol. 63, 2020.

[20] Taherdoost, H., and M. Madanchian, "Artificial Intelligence and Knowledge Management: Impacts, Benefits, and Implementation," *Computers,* Vol. 12, No. 72, 2023, p. 4, https://doi.org/10.3390/computers12040072.

[21] https://monkeylearn.com/blog/semantic-analysis.

[22] Eliyahu, S., "Overcoming Knowledge Management Obstacles with Documentation AI," *Forbes,* June 29, 2021, https://www.forbes.com/sites/forbestechcouncil/2021/06/29/overcoming-knowledge-management-obstacles-with-documentation-ai/?sh=43d200aa2b13.

[23] https://www.onixnet.com/blog/revolutionizing-document-processing-with-googles-document-ai.

[24] https://6sense.com/tech/data-science-and-machine-learning/amazontextract-vs-documentai.

5

STATED AND UNSTATED CRITERIA

5.1 STATED AND UNSTATED CRITERIA

More than two decades ago, the best business developer, and the best capture manager with whom I have ever had the opportunity to work directly, and I developed this definition of winning in the federal marketspace. After much time and many modifications, we arrived at the following succinct statement: “The winner is the prime—that is, the lead contractor for a given team—as evaluated by the customer, such as the Air Force or Department of Energy, whose approach—their proposal—is demonstrably different and superior when evaluated against stated and unstated criteria.”

The stated criteria include the RFP or other solicitation documents, questions and answers, and formal amendments to the RFP. The unstated criteria refer to those hopes, fears, biases, critical issues, and success factors that business developers learn through F2F visits with key government decision-makers. A word about “bias” in this context: think of a bias as a strong preference in one direction versus another. A case in point is the government decision-makers’ preference, or bias, for BMC Helix ITSM (IT Service Management) versus Ivanti (powered by Heat, formerly LANDESK Service Desk) for enterprise service management support. Another example is the specific agency's bias against a contractor PMO.

Every offeror has access to the stated criteria. A key difference between winning and coming in second on a single-award, competitive federal contract is your organization's awareness of the unstated criteria, and that awareness must be reflected in the words and illustrations contained in your proposal.

5.2 DEEPER INSIGHT INTO UNSTATED EVALUATION CRITERIA

On a complex solicitation that is more than $2 billion led by a civilian agency in 2023, across the 357 combined pages of the final RFP, PWS, and Data Requirements Descriptions (DRDs), KM was referenced exactly one time. One might therefore conclude that the agency did not view KM as an important approach and activity. However, the government customer indicated verbally that sharing knowledge will be critical on this specific consolidation program.

The term "learning organization" did not appear across those same 357 pages. Yet, verbally, the same government customer spoke of the contractor constantly growing, learning, and innovating. The agency's leadership actually voiced the term "learning organization."

The mindset of "work from anywhere" was reinforced verbally by agency leadership. However, this term did not appear in the RFP, PWS, or DRDs. Opening the aperture of the search, the words "virtual" and "remote" each occurred one time in the PWS. This would not lead a company to presume that working remotely was a critical dimension of program success.

The agency term of art called "mission community" was absent from the RFP, PWS, or DRDs. Yet leadership referred to this term verbally. The same was the case with the term "transparent cybersecurity."

Finally, the important outcome of business value was used in the solicitation documents precisely twice—once in the PWS and once in the DRDs. It also was discussed one time in the Applicable Documents List. Interestingly, the agency's leadership actually viewed business value as a vital overarching outcome for the program.

As is evident, the stated criteria—that is, the RFP, PWS, DRDs, Applicable Documents List, and other attachments—did not convey the importance value of critical terms and outcomes. The only way to learn these is through direct customer interaction, either virtually or F2F.

5.3 KEY QUESTIONS TO ASK GOVERNMENT CIVIL SERVANTS TO LEARN ABOUT UNSTATED CRITERIA

- General:
  - What are the top 3 characteristics that make an outstanding industry partner for working with your organization?
  - What is your vision for this contract 5 years from now?
  - What are your primary concerns regarding the success of this contract going forward (e.g., funding levels, requirements de-scoping, platform migration, or major facility move)?
  - Who are your principal stakeholders under this contract (e.g., defense and intelligence communities)?
  - For a contractor to provide you and your leadership team with peace of mind under this contract, what types of things would you want to see being done from a technical and management perspective on an ongoing basis?
  - On a day-to-day basis, how important is meeting the guidelines of the President's Management Agenda (PMA) and the requirements of the Office of Management and Budget (OMB) Exhibit 300 scorecards to your organization?
  - What types of National Institute of Standards and Technology (NIST) and/or agency-specific guidelines (e.g., DHS Security Management Directives) are most important and applied on a regular basis under this contract?
  - What types of contractor behaviors and actions build trust for you?
  - What other people within your organization do you recommend we speak with to gain additional insight into the elements of successful performance under this contract?
- Management approach:
  - What type of contractor project organization works best for you and your management team—for example, an integrated project team (IPT) or a project organization that aligns functional work areas with specific team partners?

- What value do you see in a contractor's PMO?
- What has your experience been on this contract relative to teams versus a single-vendor solution?
- For a project manager to be successful on this contract, what kinds of activities does he or she need to do on a daily basis?
- How often do you like to see and communicate with the contractor's project manager—every day, once each week, and so forth?
- What level of visibility do you expect to have into the schedule and cost aspects of this contract? Are Web-based tools important to you in this regard?
- What levels and frequency of workload fluctuations have you experienced during the current contract? What do you anticipate moving forward?
- How have you benefited in the past from contractor-provided senior advisory panels?
- How important are knowledge sharing and KM in leveraging intellectual capital to the success of this contract going forward?
- With what types of automated tools are you comfortable to monitor and control project-level and task-level costs?
- What types of organizational change management initiatives do you envision related to this contract in its upcoming period of performance?

- Technical approach:
  - What recognized standards-based processes (e.g., ISO 9001:2015, ITIL v4, CMMI) are already in place on this contract? In what ways are they being applied? How do you think that they have benefited the contract?
  - Are service level agreements (SLAs) already being used on this contract? If so, how many are established currently? What aspects of performance are these SLAs used to measure?
  - How do you see the technologies associated with this contract changing within the next year or two? For example, are you anticipating migration to a private or hybrid cloud?

  - What types of candidate technical innovations would you like to see introduced under this contract?
  - What current contract technical issues keep you up at night?
  - How important to you and the success of this contract is the ongoing technical refreshment of staff?
- Key personnel:
  - Of education and directly applicable work experience, which qualification is more important to you and to the success of this contract going forward?
- Phase-in/transition:
  - What makes a successful contract transition for you and your organization?
  - How important is retaining the current incumbent staff to the success of your organization's mission?
  - What kinds of issues concern you most during the transition phase of this contract?
- Risk management:
  - What automated risk management tools have been applied successfully to supporting this contract?
- Cost:
  - From your perspective, what are the top three things that make a best-value industry partner?

5.4 GETTING CREDIT FOR YOUR STRENGTHS—SHAPING THE FINAL RFP

At a preproposal one-on-one meeting with the Air Force in the western United States, one of the branch chiefs in attendance stated that company culture is a critical contributor to staff retention. So how can that insight be converted into strengths that can actually be evaluated by the Air Force? One way is to provide candidate language for Section L, Instructions to Offerors, along with candidate language for Section M, Evaluation Factors for Award, so the government can objectively assess corporate culture in the proposals it receives from industry.

I began by researching the definition of organizational culture: "A set of norms and values that are widely shared and strongly held throughout the organization" [1]. Next, I crafted language that the company which I was supporting built into a white paper to the Air Force.

5.5 CANDIDATE INSTRUCTIONS TO OFFERORS LANGUAGE (SECTION L)

- Prime offerors shall discuss key aspects of their organizational corporate culture, to include, at a minimum, tangible investment in workforce development, employee turnover rates during the past 3 years, diversity of the workforce, and maturity of the internal human resources function.
- Prime offerors shall characterize the established pathways by which staff professionals will be able to have access to the offeror's corporate leadership.
- Describe the process by which staff-suggested process innovations will be collected, vetted, implemented in close coordination with the Air Force, and rewarded by the prime offeror.
- Characterize the maturity level of the prime offeror's HR Department, demonstrate how the offeror will frequently update its recruiting strategies to ensure continual alignment with the Air Force's mission, describe the role of the HR Department within the corporate infrastructure, and provide evidence of current professional certifications among the offeror's human resources staff.

5.6 CANDIDATE EVALUATION FACTORS LANGUAGE (SECTION M)

- Evidence of retaining high-demand/low-density (HD/LD) personnel during the past 3 years;
- Evidence of tangible investments in workforce development;
- Evidence of low voluntary employee turnover rates during the past 3 years;
- Evidence of transparency and leadership visibility;

- Evidence of corporate recognition of staff contributions;
- Evidence of the offeror's alignment with the Society for Human Resource Management (SHRM) best practices;
- Evidence that the organization promotes a diverse and inclusive workplace where no one is disadvantaged because of their gender, race, ethnicity, sexual orientation, religion, or nationality.

The benefits to the Air Force are that the language suggested will provide the government with additional information that:

- Demonstrates strong prime-level commitment to recruit and then retain qualified staff during the entire period of performance;
- Demonstrates the prime's capacity to pivot with changes in mission focus and employee expectations to ensure that the right staff are on the job;
- Validates that the prime's critical infrastructure functions are not outsourced to a third-party service provider;
- Demonstrates that the prime is fully capable of sustaining a proactive staffing program.

At the same one-on-one meeting, the prime that I was supporting suggested that the evaluation criteria include the offeror's long-term business strategy, given the long duration of the contract. To that end, I developed language for the prime's white paper to the Air Force customer.

5.7 CANDIDATE SECTION L LANGUAGE: LONG-TERM BUSINESS STRATEGY

- Prime offerors shall include a 1-page letter on their official company letterhead in Volume I—Management Approach.
- This letter shall be signed by the executive leader of the proposing business unit, and provide a high-level projected roadmap for the prime's next 5 years of business direction in the Department of Defense (DoD) marketspace.
- This letter will not be included in the core page count.

5.8 CANDIDATE EVALUATION FACTOR LANGUAGE: LONG-TERM BUSINESS STRATEGY

- Letters will be assessed to ensure that prime offerors intend to remain within the DoD marketspace for the foreseeable future.
- The benefits to the Air Force are that the language suggested will provide the government with additional information that:
  - Provides the government with a level of assurance that the offeror is suitable for executing a contract with a period of performance of more than 12 years and that it has the corporate focus to execute against the commitments in its proposal. It also will enable the government to assess whether the company's corporate strategy aligns with the government's strategy for the contract.
  - Provides the government with the organizational structure necessary to drive consistency/common processes across task orders that will increase staffing and execution efficiencies.

Finally, the prime offered to develop evaluation criteria by which the Air Force can assess process innovations. To that end, I constructed language for the prime's white paper to the Air Force customer.

5.9 CANDIDATE SECTION L LANGUAGE: PROCESS INNOVATIONS

- Prime offerors and significant subcontractors shall provide detailed evidence in Volume I—Management Approach of their successful and collaborative introduction of process innovations on complex, multiyear federal contracts.
- Demonstrate the value of these process innovations to the government in terms of, at a minimum, enhanced quality, improved productivity, adherence to schedules, cost control and/ or avoidance, and risk mitigation.

5.10 CANDIDATE EVALUATION FACTOR LANGUAGE: PROCESS INNOVATIONS

- Evidence of the introduction and sustainment of meaningful and beneficial process improvements in close collaboration with government customers.
- Clear articulation of the value to the government of these process innovations in terms of, at a minimum, enhanced quality, improved productivity, adherence to schedules, cost control and/or avoidance, and risk mitigation.
- The benefits to the Air Force are that the language suggested will provide the Air Force with additional information that provides the government with evidence of an offeror's ability to successfully implement and manage innovative approaches.

5.11 45 STRENGTHS ARE THE NEW TARGET

Well, I found a new high bar for strengths identified in a source selection decision document. AT&T in Oakton, Virginia, received 42 strengths from the Department of Homeland Security (DHS) and the U.S. Secret Service. So, I must now revise my guidance to be 45 strengths.

The previous high number of strengths and significant strengths was found in a NASA source selection statement for the winning prime contractor on the Goddard Space Flight Center (GSFC) Space and Earth Science Data Analysis (SESDA) proposal. That total was 38.

Reference

[1] O'Reilly, C. A., and J. A. Chatman, "Culture as Social Control: Corporations, Cults, and Commitment," in B. M. Staw and L. L. Cummings (eds.), *Research in Organizational Behavior: An Annual Series of Analytical Essays and Critical Reviews,* Vol. 18, New York: Elsevier Science/JAI Press, 1996, p. 166.

6

MOVING BEYOND COMPLIANCE

6.1 PROPOSAL COMPLIANCE IS NECESSARY BUT NOT SUFFICIENT

Sustained successful proposal development is all about compliance—but I will not say that today or any day. Compliance is necessary but not sufficient to win. Focusing heavily on compliance in "proposal land" is like saying that to be successful in the National Hockey League (NHL), the primary emphasis should be on skating (see Figure 6.1). Every starting player of the 31 NHL teams is an exceptional skater, fully capable of rapid maneuvering, as well as sudden stopping and changing direction on the ice. Skating alone does not put the puck between the pipes enough times to win any given hockey game. By definition, no team can win if they do not score. Scoring every game takes passion, planning, teamwork, execution at the right time, and discipline. Add to these, expecting to win and some good luck. These same things apply to proposaling. You have to want to win more than your competitors. You have to think it, feel it, embrace it, taste it, and own it throughout the entire BD life cycle. Further, this need for total focus should tell us something very important. Most companies pay it little attention. What is the "it"? The fact that total focus should modulate your company's bid/no-bid decision-making strategy such that you do not pursue everything that appears at the

Figure 6.1 Washington Capitals' left-winger and captain Alex Ovechkin shooting the puck against the Buffalo Sabres' goalie. (Photograph © Dr. R. S. Frey.)

System for Award Management (SAM) official U.S. government website (https://sam.gov/), which remotely aligns with one of your core competencies, or a bid decision is made because the owner has a "gut" feeling about this opportunity, or a government small business specialist urges your company to bid.

Compliance fits the paint-by-numbers proposal paradigm to a tee. It reminds me of something that an active-duty, 3-tour U.S. Army combat veteran shared with me. Some officers pay inordinate attention to minute details, such as the precise height of a soldier's general service dress uniform trouser cuff above his dress tie-Oxford shoe. Yes, the Army is a professional organization, and presentation and honoring legacy are important. However, even more critical is the capacity to figure things out in-country when lots of potentially bad things are happening simultaneously. It is relatively easy to meet the mark of Army Regulation 670–1—"Wear and Appearance of Army Uniforms and Insignia." Applied to proposals, throttle down the level of energy directed toward compliance. Certainly, be compliant, but compliance should not stand as the be-all and end-all focus of the proposal development team.

6.2 FONT AND LINE SPACING: STILL CRITICAL ASPECTS OF PROPOSALING

Yes, even in 2023, companies still experience issues with font family and line spacing. Font family refers to a set of fonts that have a common design. Line spacing is also known as "leading" (rhymes with "bedding"). Traditional print shops used to place strips of lead (metal) between lines of metal type (as shown in Figure 6.2) to increase vertical space.

So let's unpack the issues surrounding font family and line spacing. The popular Times New Roman, which is a serif font, consists of roman, italic, bold, and bold italic versions of the same typeface. Originating from ancient Roman square capital letters, a serif is a small line or stroke regularly attached to the end of a larger stroke in a letter

Figure 6.2 Rule and cold metal type. (Photograph © Dr. R. S. Frey.)

or symbol within a particular font or family of fonts. However, Arial is one example of a sans-serif font (i.e., without serif). The Instructions to Offerors (Section L) in a recent federal government RFP stipulated that the font for graphics had to be 10-point Times New Roman. As an aside, one point equals 1/72nd of an inch, and is used to measure the vertical height of letters. Yet this company elected to develop graphics using Arial font in the callouts, and in far smaller than 10-point type. What this meant was that all of the graphics had to be redesigned and, in some cases, reconceptualized. This was not a prudent use of time and bid and proposal (B&P) dollars.

Also recently, another company chose to craft the technical approach for their proposal using "exactly 11.9 points (pt)" under line spacing in the paragraph group in Microsoft Word. That would have been fine, except for the fact that the RFP stipulated the use of "single space." This issue ballooned the technical approach from 16 to 22 pages. In turn, this translated to significant additional editing, tighter writing, and time invested in rework (think Lean and the removal of waste) during the proposal response life cycle that was unnecessary.

Line spacing is also critical to keep in mind for graphics. Taking the leading in graphics to the proverbial "squish" because the writers need another two lines of narrative is technically noncompliant.

The lesson is this: follow the directions in the RFP to the letter, and the number. There are federal agencies that use the kinds of missteps described above to disqualify an offeror's proposal.

6.3 STRAIGHTFORWARD COMPLIANCE CAN BE CHALLENGING

Ensuring full compliance with federal government RFPs sounds straightforward, but it can be quite complex and convoluted (see Figure 6.3). There are several circumstances that contribute to the challenges that industry faces. One is when the RFP provides instructions in Section L that are presented in paragraph format rather than with some type of numbering or alphanumeric labeling scheme. As an example, in a recent U.S. Agency for International Development (USAID) RFP, nine requirements were presented in two unnumbered paragraphs spanning 392 words. Demonstrating exacting compliance is difficult to do in this instance. In that same USAID RFP, the techni-

Figure 6.3 Reflections of the boats and water in a marina in downtown San Diego, California. (Photograph © Dr. R. S. Frey.)

cal approach section included two different lists. Unfortunately, both were labeled (1) to (5).

In a NASA RFP released in 2022—under the Mission Suitability Volume Instructions, Item B.3 Management Structure, 12 requirements were included within a single 188-word paragraph. In addition, one of those requirements was duplicated in the same paragraph.

Another circumstance that presents challenges for proposal professionals is when Section L, the Instructions to Offerors, does not align with Section M, Evaluation Factors for Award. Back to the same USAID RFP. Section L required six annexes. Yet Section M provided no language about how the government was going to evaluate those annexes.

As a third circumstance, I can point to a NASA solicitation released in 2021. This RFP stipulated the inclusion of a 4-column Requirements Traceability Matrix as part of the proposal submittal. Mapping the elements of Section L to the PWS was extremely challenging. Adding to the complexity was the fact that there were two different contract line item numbers (CLINs) associated with specific items in the PWS. Further, the language that the government used in the PWS did not align fully with the language in Section L. Building the mapping required several days of sustained effort. I know because I did it.

RFPs are complex contractual documents developed by multiple people over time. It should not come as a surprise that there are inherent inconsistencies.

6.4 BUILDING A MEANINGFUL COMPLIANCE MATRIX

Compliance, cross-reference, traceability, or cross-walk matrices, all are terms for tables that map the RFP or RFQ) requirements clearly to the specific sections within your proposal in which your company responded to a given solicitation element. The government's solicitation elements should always appear in the left column. It is industry's task to map their response to the solicitation, making it very easy for government evaluators to track full compliance.

Typically, a compliance matrix will include elements from Section C, Statement of Work (SOW); Section H, Special Contract Requirements; Section L, Instructions to Offerors; and Section M, Evaluation Factors for Award. However, in a recent complex U.S. Department of Education RFP, requirements that had to be addressed were also included in Section I, Contract Clauses and Section F, Deliveries or Performance. In Section F, there was a requirement for the offeror to provide a "written preliminary plan describing how [the offeror] will continue to perform the contractor services listed in Section C, p. 31." Additionally, this particular solicitation document also included 17 attachments.

In Department of Defense (DoD) solicitations, offerors will see a Contract Data Requirements List (CDRL) and Data Item Descriptions (DIDs). A DID is a completed document that defines the data required of a contractor and is included in a CDRL. One example of a DID entitled "Instructional Media Package" contains the preparation instructions for the content and format of that package. Your compliance matrix needs to include where in your proposal the DIDs are addressed.

In NASA procurements, you will encounter data requirement deliverables (DRDs) found within the Data Requirements List (DRL). I have seen NASA procurements include a DRL with 210 DRDs, including IT Security Plan (ITSP), Integrated Master Plan (IMP), Export Control Plan, Insight Implementation Plan, and Mission Integration and Operations Management Plan (MIOMP), as well as Quarterly Program Review (QPR) and Flight Test Readiness Review (FTRR). All of these DRDs must appear in your compliance matrix with exacting cross-mapping to your proposal.

Compliance matrices often appear in the front matter of proposal volumes. RFP instructions permitting, they may also be placed at the very end of each proposal volume as an 11 × 17-inch foldout page. The inner half of the foldout page is blank. This technique allows the government evaluators to open the foldout and refer easily to the matrix. To be sure, there are still solicitations that require hardcopy proposal submittals, as evidenced by a recent Department of Veterans Affairs opportunity that required hand delivery to Golden, Colorado.

To add value to your compliance matrix, include reference to both text sections or paragraphs as well as figures/tables contained in your proposal response. Note that, in certain cases, the response to specific RFP or RFQ elements is found in more than one proposal section or subsection. This fact must be reflected in your compliance matrix.

Whenever possible within page count limitations proscribed by the solicitation, I always encourage my customers to include a 3-column compliance matrix (see Figure 6.4). Even if there is no allowance in the final proposal volumes, I strongly suggest that organizations still build and populate this 3-column matrix. Why? Because it helps to ensure exacting compliance across the RFP, RFQ, tender, or request for solution (RFS)—the term used in the most recent commercial proposal that I supported. Importantly, it provides a mechanism to articulate strengths, benefits, and value to the customer in a detailed manner, mapped to every one of the solicitation elements.

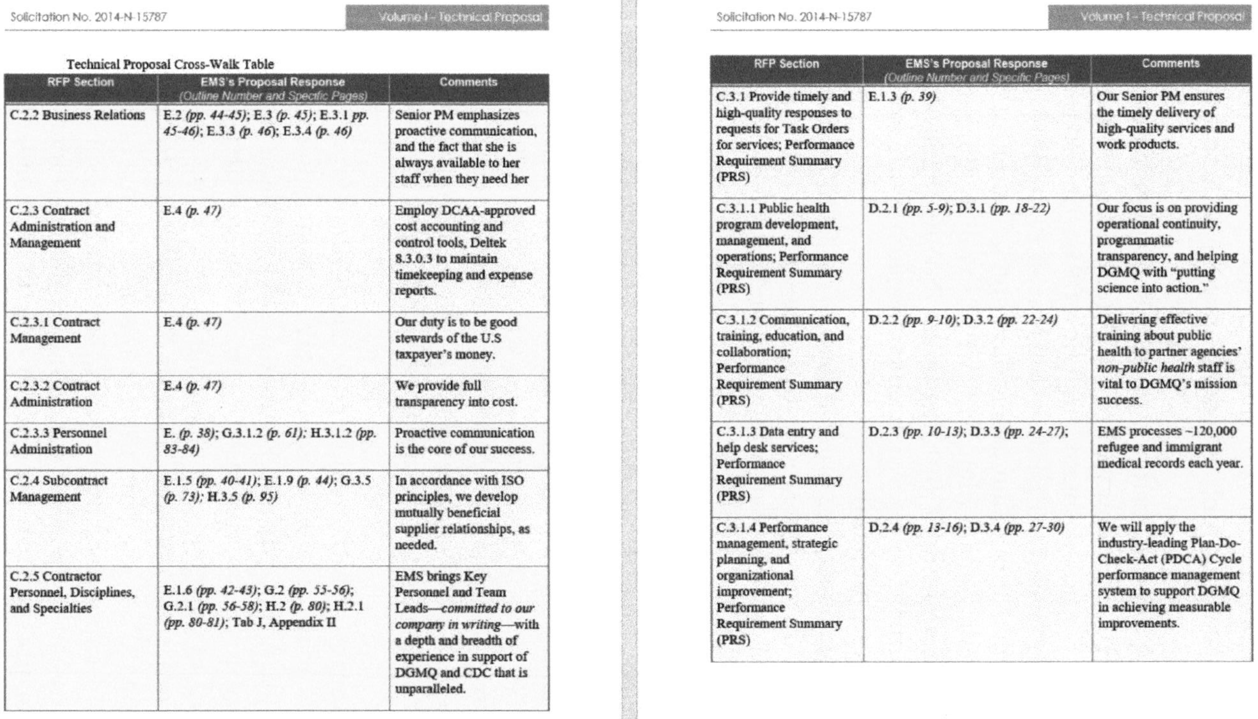

Solicitation No. 2014-N-15787 | Volume I – Technical Proposal

Technical Proposal Cross-Walk Table

RFP Section	EMS's Proposal Response (Outline Number and Specific Pages)	Comments
C.2.2 Business Relations	E.2 *(pp. 44-45)*; E.3 *(p. 45)*; E.3.1 *pp. 45-46)*; E.3.3 *(p. 46)*; E.3.4 *(p. 46)*	Senior PM emphasizes proactive communication, and the fact that she is always available to her staff when they need her
C.2.3 Contract Administration and Management	E.4 *(p. 47)*	Employ DCAA-approved cost accounting and control tools, Deltek 8.3.0.3 to maintain timekeeping and expense reports.
C.2.3.1 Contract Management	E.4 *(p. 47)*	Our duty is to be good stewards of the U.S taxpayer's money.
C.2.3.2 Contract Administration	E.4 *(p. 47)*	We provide full transparency into cost.
C.2.3.3 Personnel Administration	E. *(p. 38)*; G.3.1.2 *(p. 61)*; H.3.1.2 *(pp. 83-84)*	Proactive communication is the core of our success.
C.2.4 Subcontract Management	E.1.5 *(pp. 40-41)*; E.1.9 *(p. 44)*; G.3.5 *(p. 73)*; H.3.5 *(p. 95)*	In accordance with ISO principles, we develop mutually beneficial supplier relationships, as needed.
C.2.5 Contractor Personnel, Disciplines, and Specialties	E.1.6 *(pp. 42-43)*; G.2 *(pp. 55-56)*; G.2.1 *(pp. 56-58)*; H.2 *(p. 80)*; H.2.1 *(pp. 80-81)*; Tab J, Appendix II	EMS brings Key Personnel and Team Leads—*committed to our company in writing*—with a depth and breadth of experience in support of DGMQ and CDC that is unparalleled.

Solicitation No. 2014-N-15787 | Volume I – Technical Proposal

RFP Section	EMS's Proposal Response (Outline Number and Specific Pages)	Comments
C.3.1 Provide timely and high-quality responses to requests for Task Orders for services; Performance Requirement Summary (PRS)	E.1.3 *(p. 39)*	Our Senior PM ensures the timely delivery of high-quality services and work products.
C.3.1.1 Public health program development, management, and operations; Performance Requirement Summary (PRS)	D.2.1 *(pp. 5-9)*; D.3.1 *(pp. 18-22)*	Our focus is on providing operational continuity, programmatic transparency, and helping DGMQ with "putting science into action."
C.3.1.2 Communication, training, education, and collaboration; Performance Requirement Summary (PRS)	D.2.2 *(pp. 9-10)*; D.3.2 *(pp. 22-24)*	Delivering effective training about public health to partner agencies' *non-public health* staff is vital to DGMQ's mission success.
C.3.1.3 Data entry and help desk services; Performance Requirement Summary (PRS)	D.2.3 *(pp. 10-13)*; D.3.3 *(pp. 24-27)*;	EMS processes ~120,000 refugee and immigrant medical records each year.
C.3.1.4 Performance management, strategic planning, and organizational improvement; Performance Requirement Summary (PRS)	D.2.4 *(pp. 13-16)*; D.3.4 *(pp. 27-30)*	We will apply the industry-leading Plan-Do-Check-Act (PDCA) Cycle performance management system to support DGMQ in achieving measurable improvements.

Figure 6.4 Snapshot of a cross-walk table that I developed for a winning Centers for Disease Control and Prevention (CDC) proposal.

When Columns 1 (left) and 3 (right) are populated, reviewed, and vetted at the Blue Team, everyone can see and understand the end-to-messaging.

It is a good practice to have someone who has not been engaged in developing the proposal response to validate compliance toward the end of the proposal response life cycle. This fresh set of eyes may identify important gaps in compliance.

7

THE LOOK AND FEEL OF A WINNING PROPOSAL

7.1 WHAT DOES A WINNING PROPOSAL LOOK LIKE?

What are the look and feel of a winning proposal (see Figure 7.1)? Your proposal is a mirror in which the customer must see themselves: their mission, their strategic and business objectives, their requirements, their stakeholders, their governance structure and your understanding of that framework, their as-is environment and your perception of it, and their strategic and operational challenges and how you will address them effectively and efficiently now and over time. Make them feel at home in your proposal. Ensure that there is exacting and visible mapping to the RFP requirements through the headings and subheadings, the numbering or alphanumeric scheme, and parenthetical cross-references to the solicitation.

Use the customer's lexicon. While working with one of my customers on their recompetition of a contract at Fort Huachuca in Sierra Vista, Arizona, I interviewed the program manager. He indicated that his primary role in supporting the Army's Contracting Officer (KO) was to "clear the brush" so that the KO could focus on important mission activities. When that specific language underwent copyediting, it

Figure 7.1 San Francisco Giants' World Series trophy on display in Oracle Park in San Francisco, California. (Photograph © Dr. R. S. Frey.)

was rephrased, viewed as not being the King's English. What we need to keep in mind is that people buy from people, and people buy from people they know and trust. Use the local phrases and terms of art, and the language that you hear during site visits, Industry Day events, one-on-one meetings, and call plan execution. Bake that language into your proposal narrative. Absolutely do not let it be edited out during the proposal development response life cycle.

Your proposal must answer the "What's in it for me?" (WIFM), question from the standpoint of the government. An excellent way to do this is to align your company's validated strengths with the Evaluation Factors for Award found in Section M of the RFP. Illustrating this alignment in a table or figure is a really good idea.

Winning proposals are also characterized by relevant success stories, including intangible assets, such as accessibility evenings and weekends, an established record of highly ethical business practices, responsiveness and a strong sense of urgency, and anticipation of your customers' diverse needs.

Text must augment and support the graphics and tables. Ideally, there should be at least one graphical element on every page of your proposal—think textboxes or callout boxes, a small set of value-added icons, and drop caps.

Finally, quantitative validation rather than generalities (e.g., extensive, vast, and widespread) is critical to winning.

Next are several useful resources for increasing your organization's probability of winning federal proposals:

- Frank, J. P., *An Insider's Guide to Winning Government Contracts: Real-World Strategies, Lessons, and Recommendations,* Second Edition, St. Louis, MO: RSM Federal, 2022.
- Lohfeld, R., *10 Steps to Creating High-Scoring Proposals: A Modern Perspective on Proposal Development and What Really Matters,* Edgewater, MD: Lohfeld Consulting Group, 2017.
- SMA, *The Essential Principles of Winning Proposals,* Second Edition, SMzA, Inc., 2022.

7.2 GOOD IS THE ENEMY OF GREAT

"Good is the enemy of great" (see Figure 7.2). I saw this quote on a T-shirt several years ago. It's actually the title of the first chapter in business management author, researcher, and consultant, Jim Collins' book entitled, *Good to Great: Why Some Companies Make the Leap... and Others Don't.* It relates directly to capture management and proposal development.

I distinctly remember working on a proposal for NASA Glenn Research Center (GRC) in Cleveland when this concept really came to life. The team had developed a characterization of the ideal program manager for the opportunity. The challenge was attracting and engaging the specific individual who fit the description in all ways. The description encompassed such attributes as X years of management experience on a NASA GRC contract with at least Y staff, and relevant task-specific experience. Interviews were conducted with multiple candidates. Among the leadership of the prime contractor, there was a growing sense that the team may have to settle for the second-best or even third-best program manager candidate. However, the business

Figure 7.2 "Great" aerial view over the coast of Maui. (Photograph © Dr. R. S. Frey.)

developer with the major subcontractor—the company I worked for at the time—would not support any candidate who was not the A+ answer. Through his indefatigable persistence, the business developer convinced the prime's ownership to pursue the incumbent's program manager. In our proposal, we also built in a money-back guarantee for the program manager to work for a minimum of 1 year. There would be no bait-and-switch. Ultimately, the incumbent program manager agreed to join our team upon contract award, and she did.

This same concept also applies to decisions that federal support services contractors make regarding investments in the success of a given government program. In reviewing scores of source selection statements over a period of years, I have observed that corporate investments frequently rise to the level of a significant strength. Next are three specific examples:

- Significant strengths: "... (2) corporate investments including two choices for the IT Service Management transition to

ServiceNow, plus zero phase-in costs, and various tools and opportunities for innovation."

- Significant strength: "Corporate investments, which included phase-in cost... and investments in independent research and development (IR&D)."
- Significant strength: "Invest a significant amount of its own money in several initiatives related to process improvements, human capital development, and marketing of Langley Research Center (LaRC) research facilities."

As the prime contractor, your company should strongly consider making strategic investments in the government's program you are pursuing. It could spell the difference between good and great.

7.3 THE MOST IMPORTANT FACTORS IN WINNING

So, what are the most important factors in winning U.S. federal government proposals on a consistent basis over time? When conducting proposal training for a group of staff professionals from a company in western Ohio, I posed this question. I report what participants wrote onto the whiteboard, and also comment on select points.

1. Go/no-go assessment. A fact-based, even-handed, decision-making process focuses energies on opportunities that align with your corporate strategic plan.
2. Understanding of the customer, their buying habits, and the proposal requirements. Valid.
3. Compliance. Yes compliance is necessary, but not sufficient to win.
4. Customer relationship in advance of the final RFP. Absolutely critical to success.
5. Knowledge of the competitive landscape. Yes, it helps to drive your company's strategy.
6. Well-written proposal. Certainly.
7. Competitive price. Cost/price is always important to the government, even in best-value procurements.

8. Past performance. Recent and relevant past performance is critical. Exceptional Contractor Performance Assessment Reporting System (CPARS) reports are valuable assets as well.
9. Early corporate engagement. This is essential, along with an executive capture strategy.
10. Win themes/discriminators. Actually, no. This is industry-speak. The government evaluates industry's proposals in terms of strengths, weaknesses, and deficiencies. Focus on presenting evidence of strengths.
11. Facilities (e.g., office space, IT support). Certainly, IT support is invaluable, but the COVID pandemic demonstrated that proposals can be developed in a totally virtual environment.
12. Technical solution. Definitely.
13. Team. Yes, the scope and complexity of most procurements today cannot be addressed through a single-vendor solution.
14. Risk management approach. Good one.
15. Teamwork. Yes, across the prime and its subcontractors.
16. Proposal development process. Having a structured, repeatable, and documented process is valuable to optimize time and effectively leverage resources.
17. Proposal manager. Yes, that role is vital to orchestrate the proposal development process and ensure multidimensional communication.
18. Communication. A primary reason why projects fail is lack of effective communication. And a proposal fits the definition of project to a tee in that it is time-bound and focused on attaining a certain outcome.
19. Graphics. Yes, compelling, easily understood graphics with benefits-oriented captions help tell your company's story and prevent presenting the government evaluators with a "wall of words."
20. Targeted executive summary. Absolutely, and focused on providing evidence of how your company's strengths align with the Evaluation Factors for Award (Section M of RFPs that follow the FAR).

21. Constructive and facilitated review process. Comments such as "shred every syllable and start over" or "somebody must have gotten a dictionary for Christmas" are certainly less than helpful.
22. Attention to detail. Right on target!
23. Time management. Critical for success. Time is your most limited resource.
24. Schedule control. Important to submit a compliant, credible, compelling, and cost-competitive proposal in exact accordance with the date and times stipulated by the government. Importantly, it is the government's server clock that is the definitive decider for virtual submittals.
25. Common lexicon. Having everyone on the proposal team understand the roles and responsibilities of the other members and having a shared conception of the outcomes of various color team reviews are essential.
26. Trained proposal staff. Certainly true.

To this listing, I would add "vectored passion" (see Figure 7.3). Wanting the win more than anyone else is an intangible that must be sustained over the entire proposal life cycle.

You might think of vectoring in terms of a garden hose. Without a nozzle, the water does not go very far from the end of the hose. With a nozzle attached, the water can be focused on a home or garden project. Similarly with passion. Vectoring passion takes human energy and focuses it on a specific outcome, namely, winning a given proposal. Leadership is critical in vectoring a proposal team's collective passion.

The question that one person raised was what should be the proposal team lead's vector for winning the deal: by beating the competition, or the client's success by maximizing the value?

I recommend always focusing on maximizing the value that your team brings to the client—which is also aligned to the solicitation's Evaluation Factors for Award. Personally, I do not focus a great deal of attention on the competition, I have seen many instances in which a relatively unknown team or company comes out of the blue and wins the opportunity. Focus on what you and your team can control.

Figure 7.3 The Lockheed A-12 was a high-altitude reconnaissance aircraft. It was the precursor of the YF-12 and SR-71 Blackbird. The thrust-vectoring nozzle supported maneuverability. (Photograph © Dr. R. S. Frey.)

7.4 STAYING SHARP—BUSINESS ACQUISITION AS A FORMAL PROCESS

Dedicated effort in accordance with a well-defined, yet flexible plan; broad-based and in-depth knowledge of your customers and competitors; superlative performance on past and present projects; and a formalized communication and knowledge-sharing network all contribute to successful proposals. Let's expand upon this listing.

In an extremely important quantitative study conducted by Price Waterhouse in the 1990s (before its merger with Coopers & Lybrand to become PwC), it was determined that companies that exhibit superior performance as measured by competitive contract awards managed "their business acquisition as a formal, disciplined process. These companies view business acquisition as a structured set of interrelated activities to win contract awards. The superior performers continuously improve the methods they use to pursue opportunities" [1]. In effect, innovation should stand as a core business strategy, as noted by Farida and Setiawan [2]. Certainly, there are times when continuous improvement or the formation of new competitive advantages necessitates some level of organizational change, which can be challenging even in small organizations. However, I have lived the reality that small businesses can be much more agile and pioneering than large corporations. Importantly, small businesses can also win federal full-and-open competitive acquisitions against large, publicly traded corporations.

Writing in *Forbes,* Scott Pollack, cofounder and CEO of Firneo, succinctly and accurately characterized business development as "the creation of long-term value for an organization from customers, markets, and relationships" [3]. Effective and sustained "business acquisition" in O'Guin's lexicon and "business development" in Pollack's terminology requires ongoing environmental awareness of opportunities and threats, and associated organizational adjustment.

Figure 7.4 Light from a laser pointer "paints" red onto a Speedbor wood-boring drill bit. (Photograph © Dr. R. S. Frey.)

Small businesses are well-advised to conduct semiannual strengths, weaknesses, opportunities, and threats (SWOT) analysis sessions that include a broad cross-section of business development, capture management, proposal development, infrastructure, and operations managers, as well as executive leadership.

In addition, small businesses with a PMO can leverage this organization to share business development and proposal development information, lessons learned, and best practices across the operations staff. The PMO can also train and assist operations staff in collecting quality, schedule, cost, and risk mitigation metrics and success stories from the contracts on which they are working.

The important action is to keep improving how your firm pursues and captures new and recompete business with the federal government. Document this "living" process, and be sure to codify all of the interfaces and hand-offs. Keep your business development process sharp (see Figure 7.4).

References

[1] O'Guin, M., "Competitive Intelligence and Superior Business Performance: A Strategic Benchmarking Study," *Competitive Intelligence Review,* Vol. 5, 1994.

[2] Farida, I., and D. Setiawan, "Business Strategies and Competitive Advantage: The Role of Performance and Innovation," *Journal of Open Innovation: Technology, Market, and Complexity,* Vol. 8, No. 3, 2022, p. 202.

[3] Pollack, S., "What, Exactly, Is Business Development?" *Forbes,* March 21, 2012, https://www.forbes.com/sites/scottpollack/2012/03/21/what-exactly-is-business-development/?sh=618d35607fdb.

8

HOPSCOTCHING THROUGH PROPOSAL-LAND

8.1 PROPOSAL INTEGRATION MAP

Graphics stand as a critical proposal development tool. Visualizing the connections across the elements of a federal government RFP or RFQ in a graphic enables members of the proposal team to see how components of the solicitation document relate to each other. I developed the integration map shown in Figure 8.1 when helping one of my customers to build a proposal response to a NASA solicitation that was fully connected and consistent across volumes.

Past performance citations must clearly show evidence of how and for which federal agency your organization implemented the specific technical and programmatic approaches that you are now offering in the particular proposal. For example, if your overarching approach includes applying Scaled Agile Framework (SAFe®) organizational and workflow patterns for implementing Agile practices, you will want to ensure that your past performance volume includes validation of how your organization has enhanced quality, increased productivity through measurable improvements, reduced resource requirements, drove down rework, and provided continuous delivery through SAFe for other customers. In addition, demonstrate in your management approach the shift from command-and-control-style

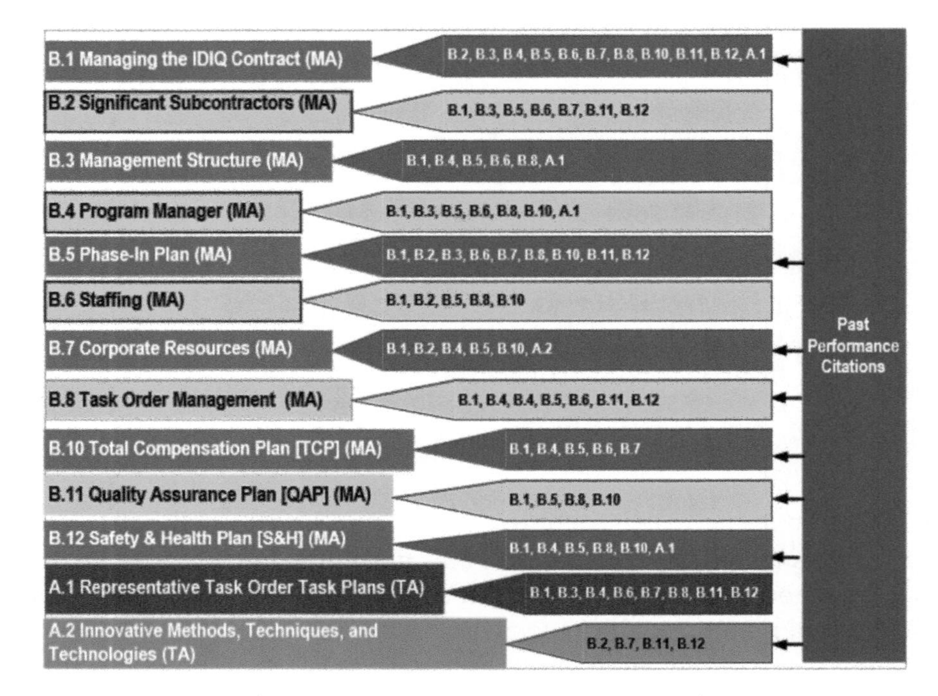

Figure 8.1 Proposal integration map supports connectivity and consistency.

program management to one that places value on coaching and allowing teams to make informed choices. Your communication plan will need to take on accentuated importance because of the criticality of transparency and collaboration under the SAFe model.

To underscore, every part of your proposal must connect with and be consistent with every other part. The integration map, which I conceptualized and developed, helps you to visualize and validate all of these connect points. In this particular example taken from a NASA RFP, the left side of Figure 8.1 captures the 12 management approach (MA) elements along with the two technical approach (TA) elements from Section L, Instructions to Offerors. In the center, all of the cross-connections are illustrated. For instance, B.7, Corporate Resources, is related to B.1, B.2, B.4, B.5, B.10, and A.2. Two cases in point—Corporate Resources must be considered during the Phase-In Plan, B.5, as well as in conjunction with Innovative Methods, Techniques, or Technologies, A.2. Importantly, past performance citations in the purple column on the right must show evidence of the company applying specific corporate resources to perform the work. This evidence should be accompanied by quantitative outcomes and benefits to the government customers.

8.2 FRAMING AND PROPOSAL DEVELOPMENT

Why propose a project manager, deputy project manager, and three task leaders to manage a project on which the customer is used to interacting with one "working" project manager and two "working" task leaders? Why introduce a six-company team to perform on a program that has been awarded historically to a contractor that offers a streamlined, two-vendor solution? Why design and build your proposal documents in a manner that does not adhere to the structure, nomenclature, and sequence provided in Sections L, M, and C of the federal government's final RFP—an RFP that the government spent months developing and reviewing? Why frame the benefits of your organization's solution in terms of exacting schedule control when decreased total cost of ownership (TCO) is the critical customer success factor? Increased proposal success follows in part from framing your approaches to connect with the known.

The framing theory suggests that how something is presented (i.e., the frame) influences the choices people make. Framing is conceptually connected to the underlying psychological processes that people use to examine information, make judgments, and draw inferences. Significantly, framing activates existing beliefs and thoughts, rather than adding something new to an individual's beliefs about a given issue. It is helpful to present your proposal story in a mode familiar to most evaluators.

One of the key questions that a company must ask when pursuing a bid opportunity to provide services, products, or knowledge sets to the U.S. federal government is: Who must we be to win? Initially, this question might appear confusing. So let us unravel it to see why it is so important. For a given bid opportunity, emphasizing (that is, giving salience, academically speaking) your organization's woman-owned business status and core competency in telecommunications support may well be the ideal picture that you want to paint. On another bid opportunity, your firm's financial stability, core competency in software engineering, and being a known NASA contractor may constitute the image that you want to frame and then project to your customer at multiple points and in multiple ways (e.g., narrative, graphics, cover art) throughout your proposal. Note that the majority of the descriptive elements that comprise your organization's overall image are not employed when answering the question: Who must we be to win? Most are omitted.

Deciding what information to present (inclusion), what points to emphasize (salience), and what information to omit (exclusion) is a critical process in framing exactly what image of your company you want to project to your customer. In his ballad "Against the Wind," guitarist Bob Seger sang about, "What to leave in, what to leave out." These song lyrics are directly analogous to framing as it relates to proposals. We as proposal professionals have to make very conscious choices about what to leave in—that is, include within the frames that we present to the government—and what to leave out.

American journalist and political commentator Walter Lippmann, who wrote the influential book *Public Opinion* in 1922, asserted that people act based on "the pictures inside our heads" rather than on objective reality. George Lakoff, Professor Emeritus of linguistics and cognitive science at the University of California (UC) Berkeley, wrote that we understand words on the basis of a "mental map." In general, people create stories to organize their experiences, create order, gain perspectives, and make evaluations.

Frames (see Figure 8.2) offer proposal professionals a powerful tool for aligning proposal solutions with customers' preexisting sentiments and mental pictures. Try applying them on your next proposal; they work.

Next is a list of resources related to this section:

- Chyi, H. I., and M. McCombs, "Media Salience and the Process of Framing: Coverage of the Columbine School Shootings," *Journalism & Mass Communication Quarterly,* Vol. 81, Spring 2004, pp. 22–35.
- Escalas, J. E., "Narrative Processing: Building Consumer Connections to Brands," *Journal of Consumer Psychology,* Vol. 14, 2004, pp. 168–180.
- Gray, B., "Strong Opposition: Frame-Based Resistance to Collaboration," *Journal of Community & Applied Social Psychology,* Vol. 14, 2004, pp. 166–176.

Figure 8.2 Tracing red laser light on a picture frame. (Photograph © Dr. R. S. Frey.)

- Hallahan, K., "Seven Models of Framing: Implications for Public Relations," *Journal of Public Relations Research,* Vol. 11, 1999, pp. 205–242.
- Ibarra, H., and K. Lineback, "What's Your Story?" *Harvard Business Review,* Vol. 83, January 2005, pp. 64–71.
- Nelson, T. E., Z. M. Oxley, and R. A. Clawson, "Toward a Psychology of Framing Effects," *Political Behavior,* Vol. 19, 1997, pp. 221–246.

8.3 MAKING YOUR BUSINESS PROCESSES STAND OUT

Naming and then trademarking (™), service marking (℠), or copyrighting (©) your company's business processes has been found to increase the likelihood of those processes rising to the level of a strength in proposals to the U.S. federal government. I know this to be the case based upon my detailed analysis of scores of source selection statements (SSS) and source selection decision documents (SSDD) from across civilian and defense agencies. Leveraging the findings of my analysis, certain business processes appear with some regularity in source selection documents. These include recruiting, training/cross-training, onboarding, and employee incentive programs. Importantly, there are metrics that your company should capture on an ongoing basis to substantiate the effectiveness of these business processes. Collecting this quantitative evidence will increase the probability that the business processes are determined to be of tangible value to the government evaluators and therefore documented as a strength in the source selection document.

- *Recruiting process:* Relevant metrics—number of high-demand/low-density positions filled per unit time, time to fill (how efficient a recruiting team is at filling an empty position).
- *Training/cross-training process:* Relevant metrics—training return on investment (ROI), operational efficiency.
- *Onboarding process:* Relevant metrics—time to start (efficiency measurement of the entire recruiting process from the date a job becomes vacant until the date a new hire finishes the onboarding process and starts the new position), time to productivity (TTP) based upon key performance indicators (KPIs) for each given position, and voluntary turnover rate.
- *Incentive programs:* Relevant metrics—number of qualified referrals received, actionable innovative concepts developed and shared.

8.4 COMMUNICATION AS A CRITICAL SUCCESS FACTOR IN PROPOSAL DEVELOPMENT

“There [is a] strong relation[ship] between communications and trust in determining virtual teams’ success” [1]. In 2022 and 2023, I supported a multibillion-dollar U.S. federal government civilian agency proposal effort over the course of 7 months. This time period encompassed preproposal engagement, source selection training, proposal development and review, and oral presentation slide and script development and presenter coaching. My customer was a prestigious global services firm with many billions of dollars in annual revenue as of 2022.

Three situations surfaced after nearly 7 months of working together virtually. To put a finer point on it, the many in-house staff professionals and outside consultant contributors were separated by space and time and were linked by shared information and telecommunications [2, p. 2]. Quite suddenly, the level of top-down communication slowed to a slight trickle. The response to upward communication decreased markedly. Proposal team members still shouldered responsibility, but with minimal allowable self-initiative. Everyone walked on virtual eggshells.

Having experienced this suboptimal situation first-hand, I began to research the academic underpinnings of high-performance virtual teams.

Dr. Timothy Brandon presented three key elements associated with mitigating the potential risks associated with virtual teams: (1) celebrate achievements, (2) establish and foster a collaborative climate based on trust, and (3) ensure that leadership is attuned to the unique dimensions of a virtual environment [2, p. 11]. In the proposal development situation described above, there was very little celebration of anything. The opportunities for senior leadership to highlight recent accomplishments and draw positive attention to specific team members or working groups were largely missed.

During the course of the 7 months, morale never rose above a modest level. Team empowerment is defined as "increased task motivation that is due to team members' collective, positive assessments of their organizational tasks" [3]. This was not a critical success factor for the pursuit leadership. In addition, there was minimal video interaction. In most Microsoft Teams meetings, the majority of participants appeared as circles with their initials. However, an empirical study recognized that "the ability to see each other's faces led video participants to perceive their teammates as more reliable" [4]. Research shows that body language, subtle voice inflections, and facial expressions, which are notably more difficult to convey via communication technology, are essential to the development of trust [5, 6]. This is certainly important to establish and sustain. Furthermore, positive, people-focused leadership was definitely an issue. Effective leaders should provide "guidance, motivation, and focus" [7].

In describing high-performance virtual teams, Dr. Brandon highlights that "[e]everyone trusts each other" and "[e]everyone can raise issues and propose a resolution" [2, p. 14]. Trust was not at all a connector with this particular proposal team. "The quality of effective collaboration can frequently be strengthened with the help of trust" [8]. Instead, rigid hierarchical structure and centralized decision-making were underscored and enforced.

Both rich discourse (i.e., containing social information and nonverbal cues as well as words, typically provided by F2F communication), and spontaneous, informal communication have been identified as key to preventing conflict and improving trust in virtual teams.

Mechanisms for supporting informal communication (e.g., chance encounters) are similarly necessary [9].

Fundamentally, preproposal activities and the entire proposal response life cycle could have been enhanced significantly had positive recognition, trust-based collaboration, and effective leadership been integral parts of the prime contractor's overarching approach. Hopefully the outcome will be worth the arduous journey.

8.5 FEDERAL PROPOSAL DEVELOPMENT—COMPREHENSIVE RISK MANAGEMENT APPROACH

Risk management and mitigation stand as critical elements to address in competitive proposals to U.S. federal government agencies. One such requirement appeared in a $250 million NASA RFP in February 2022: "The offeror shall identify the most significant potential risks under this contract and also describe the risk management techniques that will be used to manage identified risks during contract performance."

What I see so many times in this regard when reviewing proposals for small businesses, as well as *Fortune* 500 companies, is a generic combination of an unmodified Google image (the often-seen 5-step cycle with "Risk Management Process" in the center) and a table with ill-defined colorimetric designations that all show green once the proposed mitigation strategies have been applied.

How might risk management be approached differently to increase the probability that your organization's risk solution will be identified as a strength by the government's Source Evaluation Board (SEB)? You might begin by selecting a relevant risk architecture, such as the National Institute of Standards and Technology (NIST) Risk Management Framework (RMF), or NASA's 5 × 5 Risk Matrix. Invest the time to have a graphic artist customize these foundational architectures to the solicitation at hand. Align the risk architecture diagram to the Quality Assurance Surveillance Plan (QASP) if it is provided in the solicitation documents.

Importantly, initial risk mitigation steps, countermeasures, or compensating controls still often result in what is called residual risk. Chaired by the DoD, the Committee on National Security Systems (CNSS) defines "residual risk" as the "[p]ortion of risk remaining after security measures have been applied." This term also appears in very recent NIST Special Publications (SPs), including "Guide to Operational Technology (OT) Security" NIST SP 800-82r3 ipd (April 2022) and "Using Business Impact Analysis to Inform Risk Prioritization and Response" (June 2022). Your organization ought to consider discussing residual risks and also depicting residual risks through a data-driven illustration. This approach makes for a far more powerful and engaging risk management section in your proposal.

8.6 FEDERAL GOVERNMENT PROPOSALS—RESPONSIBLE CONTRACTOR

Recently, a $200 million Air Force procurement issued through a General Services Administration (GSA) RFQ included the following language under EVALUATION PROCESS/BASIS OF AWARD: "The [Blanket Purchase Agreement] BPA will be established with the responsible Contractor whose quote conforms to the requirements outlined in this RFQ and is most advantageous to the Government based on the best value determination." The key term is "responsible"; it seems like solicitation boilerplate, and many contractors read right past it, but they do so to their detriment.

Another example is this one from a recent major DoD solicitation: "The contracts will be awarded to the Offerors who are deemed responsible in accordance with the FAR, whose proposals conform to the solicitation requirements (including all stated terms, conditions, representations, certifications, and all other information required by Section L of this solicitation), and is judged, based on the evaluation factors to represent the best value to the Government, considering both price and non-price factors." In this instance, "responsible" is linked with the FAR and indeed it is. To put a finer point on things,

FAR 9.104 contains the 6-part definition of "responsible" in the context of source selection and acquisition.

NASA has become increasingly focused on the term "responsible," as evidenced in a very complex 2022 solicitation that states: "A contract may only be awarded to an Offeror who is determined to have an adequate accounting system and determined responsible in accordance with FAR 9.104." However, note that the FAR definition of "responsible" extends well beyond financial assets and resources to encompass performance schedule, performance record, record of integrity and business ethics, quality assurance measures, and safety programs, among other elements.

Contractors must ensure that they address this critical requirement in the appropriate volume of their proposal submission, whether that be the cost/price or offer volume. Make certain that this seemingly minor detail is included in the master list of action items for your proposal, and in the compliance matrix. Do not let this critical requirement fall through the cracks.

8.7 TAKE CONTRACTOR RESPONSIBILITY DETERMINATIONS SERIOUSLY

In accordance with FAR Part 9.104-6, when making a contractor responsibility determination (CRD), the contracting officer (CO) must consider all the information available through the Federal Awardee Performance and Integrity Information System (FAPIIS) with regard to the offeror and any immediate owner, predecessor, or subsidiary identified for that offeror in FAPIIS, as well as other past performance information on the offeror. The FAPIIS database was created because the U.S. government realized that COs would benefit from a single system from which to add and pull information regarding responsibility [10]. FAPIIS.gov integrity records are now called Responsibility/Qualification (R/Q) under SAM.gov. This information provides a single view of exclusions (e.g., prohibitions, restrictions, and voluntary exclusions) and proceedings (e.g., subject of a criminal, civil, and/or administrative proceeding at the federal or state level in connection with a federal award that resulted in a conviction or finding of fault or liability). Other responsibility information encompasses defective pricing (FAR 15.407), DoD Determinations of Contractor Fault, and subcontractor payment issues (FAR 42.1503(h)(1)(v)).

Well in advance of submitting a proposal, make certain that your company's R/Q records at SAM.gov are spotless. Know what information regarding exclusions and proceedings is documented there.

8.8 ORAL PRESENTATIONS

8.8.1 Best Practices for Virtual Oral Presentations

During the depths of the COVID-19 pandemic, and extending even to today, the U.S. federal government elected to conduct oral presentations in a virtual manner. I have served as an oral presentation coach for proposals to a cross-section of federal agencies, including the U.S. Department of Health and Human Services (DHHS), General Services Administration, and NASA. Three critical dimensions emerged in the process of coaching: (1) technology and techniques, (2) content, and (3) delivery.

Insofar as (1), technology and techniques, presenters need to practice and learn to use the collaborative tool, whether that be Microsoft Teams, Zoom, WebEx, or something else. Even a minor item such as signing in with one's name and appropriate position title for the program being pursued helps to establish a sense of team. I suggest using a professional clip-on microphone (no headphones) and a ring light for proper illumination of your face, elevating your computer screen to eye level, and looking directly into your camera. The camera should have at least 1,080 pixels or even 4K (4,000 pixels) of horizontal display resolution.

Regarding (2), content, minimize the amount of text on each slide. Less is actually more, including less data. Build in compelling visuals. Include visual guideposts ("bread crumbs") in the presentation. Slides need to be decipherable in 3 to 5 seconds. Avoid producing cognitive overload among the evaluators—one main idea per slide, plus one image. Be sure to tell at least one relevant story based upon your own professional experience. Focus on one goal: what do I want the virtual audience of government evaluators to know, feel, and do? Ensure that there is a clear call to action.

Now to (3), delivery. As a presenter, keep your level of energy up, not manic, but upbeat. Be in the moment, and convey passion and conviction. Vary your spoken volume, and add emphasis through voice inflection. Sit up straight in your chair, and be sure to smile. Wear professional business attire.

Select a primary and secondary timekeeper, and communicate via group texts to keep everyone on track time-wise. Be sure to keep your phone out of sight on your desk or table.

Importantly, ensure that the background in the room in which you are presenting looks neutral and professional, and ideally, uniform. One idea by which you can visually brand your organization on the videoconference is for each presenter to have a reasonably priced retractable custom mesh banner beside or behind them. You send the important message that you are indeed one team. This gives the government evaluators something visual by which to identify and remember you. When the government is controlling the collaborative platform, your company may not be able to add an electronic background of your choosing.

Next are two recent books that you will find helpful as you prepare for an oral presentation before a federal government customer.

- Stewart, J. P., and D. Fulop, *Mastering the Art of Oral Presentations: Winning Orals, Speeches, and Stand-Up Presentations,* New York: Wiley, 2019.
- Pinvidic, B., *The 3-Minute Rule: Say Less to Get More from Any Pitch or Presentation,* New York: Portfolio/Penguin, 2019.

8.8.2 Selecting the Optimal Oral Presenters

When the U.S. federal government includes oral presentations as part of the proposal submission, you should view this as a great opportunity to connect with your government customer. The primary goal of each presenter should be to build trust with the evaluators. Orals can be delivered in a number of different ways: (1) F2F at the government customer's location, (2) virtually through Microsoft Teams or some other virtual collaboration platform, or (3) in video format (e.g., MP4 file).

So, how should the prime contractor select the individuals to serve as presenters? If the government allows subcontractors to present, they should be considered as well. Your company may choose to have an individual from one of your small business partners serve as a presenter to show inclusion of a specific socioeconomic category on your team, or you may want to draw upon one of your large business teammates to present to demonstrate additional technical depth and geographic reach. Ensure that your company knows whether presenters' names must appear on the organization chart in the written proposal and whether presenters must be engaged directly in contract performance upon contract award. Many times, the government wants to see and "interview" the staff with whom they will be working on a daily basis.

In terms of the attributes and characteristics of the presenters, you might consider the following:

- Experience working in support of the specific government organization. Does the person understand the government customer's mission, strategic goals and objectives, operational cadence, and stakeholder community?
- Relevant certifications (e.g., AWS Certified Solutions Architect, Microsoft Azure Fundamentals, Cisco Certified Network Professional (CCNP), and IBM Artificial Intelligence (AI) Engineering Professional Certificate) and affiliations (e.g., Association for Computing Machinery (ACM), The Computing Technology Industry Association (CompTIA), Association of Information Technology Professionals (AITP), and Network Professional Association (NPA)).
- Appropriate representation of your company's commitment to Diversity, Equity, Inclusion, and Accessibility (DEI&A).
- Availability, given the presenter's other billable work.

Know that government evaluators are just like you. They will probably search for information about your presenters through LinkedIn and Facebook. Be certain that each one of your presenters has updated his or her LinkedIn and Facebook profiles. As the prime, help members of your own company as well as your subcontractors who will be presenters to update their profiles to ensure that their educational achievements, professional certifications, industry affiliations, and relevant publications are current. In addition, ensure that all posted photographs and videos are professional and in good taste.

8.9 INQUIRING MINDS

Below is a series of disparate but important questions posed to me by participants in my proposal training seminars convened in Ohio and Maryland.

Q: How would you describe a method for data management?

A: Build strong file naming and cataloging conventions for proposal-related knowledge artifacts. Tag artifacts (e.g., narrative, graphics, icons, photos) with informative metadata made consistent through a concise, Company ABC-developed lexical dictionary. Example: Help Desk, End-User Support, and Deskside Support should all yield the same query results.

Q: My question relates to Company ABC's transitioning from a reactive proposal development mentality to a thoughtful and well-planned proposal life-cycle methodology. I use the terms mentality and methodology deliberately—my experience has been that we dive into new ideas/methods with vigor but lose momentum when we all get busy with other priorities like project work.

A: That is where strong and communicative proposal management is so valuable. Daily 15-minute tag-up meetings with full participation and clear agendas are helpful. Identify progress, completed actions, and impediments. Leveraging a Kanban Board (To Do, In-Progress, Done) is extremely useful to visualize status and progress.

Q: What are the compelling reasons to use present tense when writing a proposal?

A: *The Tongue and Quill* (AFH 33-337) [11], which is used by the Army, Air Force, Marines, Navy, and Coast Guard, stipulates to use verbs in the present tense [11, p. 79]. The National Archives and Records Administration (NARA) *Writing Style Guide* (last updated on January 13, 2023) [12, p. 32] asserted that: "[t]he simplest and strongest form of a verb is present tense. Using the present tense makes your document more direct and forceful and less complicated."

Q: What are the pitfalls, if any, to using a previous successful proposal as a template for the same client?

A: Company ABC should always build a new annotated outline for each proposal that your company pursues. Then and only then should artifacts from previous proposals—including successful ones—be considered for the response to the new proposal. Note that some unsuccessful proposals may also have had strong sections, such as phase-in or management. These should be considered as well. The important point here is not to simply reuse a previous successful proposal for the same agency for a new pursuit.

Q: How do you determine what the "voice" is or should be when utilizing multiple contributors?

A: Build a comprehensive list of writing conventions that are specific to the RFP, the customer, and Company ABC's corporate direction. For example, decide if you will use "The Company ABC Team" or "Team Company ABC."

Q: I am curious, as someone who graduated in 2020 and is newer to the workforce and may not have as much expertise/subject matter experience, how to toe the line of describing one's own (or company's) capabilities in a confident way, but without overpromising. Truly, my question is: how do you make a proposal stand out? What truly makes a certain proposal a winner?

A: Think of the proposal as the final exam in the business development life cycle. Winning is contingent on multiple factors, such as: (1) does the government customer know and like Company ABC—people buy from people whom they know and trust; (2) has Company ABC pre-sold the customer regarding your company's solutions; (3) will Company ABC's federal customers provide stellar responses to Past Performance Questionnaires (PPQs); (4) is your proposal compliant (easily traceable to the RFP), compelling, and credible; (5) is Company ABC offering a competitive price that will not be "should-costed" up; and (6) does your proposal convey evidence of strengths that align with the government's Evaluation Factors for Award (Section M)?

Q: Do customers typically know their problems, shortcomings, and areas for opportunity, or does proposal writing require some back and forth with the customer to get at the meat for a better curated proposal? Do those opportunities for community typically exist when finding proposals to submit?

A: The federal government issues solicitations based upon a set of requirements. However, this does not mean that the evaluators for a specific proposal fully understand the scope of their organization's problems or potential opportunities. That is where business development is so valuable to meet with government decision-makers to engage in dialogue to help shape how they perceive their issues, and how Company ABC can bring them value for their mission, objectives, technology roadmap, and so forth.

Q: After a document is stored in a knowledge repository, when should it be archived and no longer considered for proposals? Is there a typical life span?

A: The answer is highly dependent upon what the document is. For example, if it is a Company ABC corporate overview, its shelf life may be measured in just a few months as new contracts are awarded and the number of staff increases. If it is a risk management plan, it may be useful for more than 1 year. Résumés should be updated at least once each year with new certifications, achievements, photos, and projects.

Q: How well do you think ChatGPT would draft parts of a proposal?

A: Actually, a longtime colleague and friend of mine in the federal marketspace conducted a test using ChatGPT to draft part of a response to a NASA RFP on which both he and I had worked. Both of us concluded that the narrative generated by the AI tool constituted a reasonable first draft. Not a winning proposal, but a decent start. Suggest that Company ABC leverage this or other AI tools as a complementary writing resource—not to replace the human writer, but to accelerate the drafting process.

Q: What are your thoughts on using a less formal writing style in proposals (for example, saying "you" instead of addressing "the client)?

A: I am a huge proponent and practitioner of humanizing proposal writing. Once the parties are identified clearly, recommend that Company ABC use "you" or "your" when referring to the government, and "we" and "us" when referring to Company ABC. Work to break down the "us" versus "them" barrier.

Q: Why are discriminators not as important as strengths?

A: "Discriminators" are part of industry's lexicon. The federal government does not think in those terms, nor do they evaluate proposals using that terminology. This is evidenced from source selection statements, the DoD's acquisition guidance, and official debriefings from such agencies as the U.S. Department of Justice. In the latter, a strength is defined as: "A strong attribute or quality of particular worth or utility; an inherent asset."

Q: I am still confused about the distinction between a proposal manager and a capture manager. Can you clarify?

A: The capture manager orchestrates the development of the solution set (i.e., the choices that an organization makes with regard to technical and management approaches, processes, tools, and techniques); manages the resources across the team; writes the Win Strategy white paper; and "owns" the bid. The proposal manager ensures that the solutions developed and vetted during the capture process are integrated into the proposal volumes appropriately. This individual attends to the myriad of details associated with issuing data calls to teammates, scheduling reviews, proposal production, document configuration management, and staying current with government-issued amendments and questions and answers (Q&As).

Q: How might we facilitate the writing of the nontechnical aspects of a proposal?

A: Many of these sections can be written and illustrated in advance, and then tailored. For example, your company's risk management plan, subcontractor management plan, Organizational Conflict of Interest (OCI) plan, training and cross-training approaches, and retention successes can be crafted before you have an RFP staring you in the face. That is because they remain largely the same across multiple bids.

Q: Can a proposal be "too technical"? How technical are evaluators, or does that depend on the customer?

A: Be aware that your proposal needs to resonate with several different government audiences—technical SMEs, contractual staff, legal analysts, and perhaps ex officio evaluation panel members.

Q: Is it a good or a bad idea to go beyond the expectations of the contract—either in the proposal itself, or after winning a proposal?

A: The very definition of a strength is an aspect of an offeror's proposal that has merit or exceeds specified performance or capability requirements in a way that will be advantageous to the government during contract performance. That being said, during contract execution, your company's program manager will need to be mindful of scope creep.

Q: What stands out for me whenever I look at any government document (SOW, RFP, PWS) is the inferiority of the grammar, punctuation, and sentence structure. How important are those things to them as they review proposals and other documents we send?

A: Think of your company's proposals as actual deliverables to your federal customers. Ensure that the documents are well-written, contain zero typos or grammatical errors, and have no run-on or overly lengthy sentences.

Q: In your personal experience, what are a few big differentiators in failed proposals versus successful proposals?

A: The #1 and #2 nonprice reasons why companies of all sizes lose competitive federal proposals are: (1) lack of demonstrated understanding of the customers environment, and (2) lack of a clearly articulated approach for delivering the required services. Importantly, past performance does not equal approach. It is not wise to state that our company is implementing a given approach for the U.S. Department of Agriculture, and we will do the same for you, Mr. and Ms. Air Force evaluator.

References

[1] Yusif, B. N. M., "Communications and Trust is a Key Factor to Success in Virtual Teams Collaborations," *International Journal of Business and Technopreneurship,* Vol. 2, No. 3, 2012, p. 407.

[2] Brandon, T. P., "Building High Performance Virtual Teams in a Culturally Diverse Environment," Project Management Institute, Westchester, NY, March 24, 2023, p. 2.

[3] Kirkman, B. L., et al., "The Impact of Team Empowerment on Virtual Team Performance: The Moderating Role of Face-to-Face Interaction," *Academy of Management Journal,* Vol. 47, No. 2, 2004, p. 176.

[4] Baker, A. L., "Communication and Trust in Virtual and Face-to-Face Teams," Unpublished Doctoral Dissertation, Embry-Riddle Aeronautical University, 2018, p. 90.

[5] Bos N., et al., "Effects of Four Computer-Mediated Communications Channels on Trust Development," in L. Terveen, et al. (eds.), *Proceedings of the Conference on Human Factors in Computing Systems (CHI'02),* 2002, pp. 135–140.

[6] Olson, J. S., and G. M. Olson, "Bridging Distance: Empirical Studies of Distributed Teams," in D. Galletta and P. Zhang (eds.), *Human-Computer Interaction in Management Information Systems: Volume II: Applications,* M. E. Sharpe, Inc., 2006.

[7] Indeed Editorial Team, "7 Characteristics of Effective Teams (with Benefits & Tips)," *Career Guide,* February 27, 2023, https://www.indeed.com/career-advice/career-development/characteristics-of-effective-teams#:~.

[8] Thampi, D., R. Das, and M. V. Mahesh, "Trust in Virtual Teams and Team Effectiveness," *International Journal of Innovative Research in Technology (IJIRT),* Vol. 9, No. 8, 2023, p. 253.

[9] Morrison-Smith, S., and J. Ruiz, "Challenges and Barriers in Virtual Teams: A Literature Review," *SN Applied Sciences,* Vol. 2, May 2020.

[10] Naylor, A. M., *Improving the Contractor Responsibility Determination Process,* MBA Professional Project, Naval Postgraduate School, Monterey, CA, 2020.

[11] *The Tongue and Quill,* Air Force Handbook 33-337, 2016.

[12] The National Archives and Records Administration, *NARA Writing Style Guide,* last updated January 13, 2023.

9

THE IMPORTANCE OF UNDERSTANDING: SECTIONS L AND M

9.1 PAINT A PICTURE OF GENUINE UNDERSTANDING IN YOUR PROPOSALS

Invariably, federal government competitive solicitations include a requirement to demonstrate understanding, particularly within the technical section of the proposal. Whether the solicitation is an RFQ or RFP, the language that the specific government agency uses to request industry's understanding is very much alike.

According to an Air Force RFQ from 2022: "Quotes must demonstrate a clear understanding of the nature and scope of the work required. Failure to provide a realistic, reasonable, and complete quote may reflect a lack of understanding of the requirements and may result in your quote receiving no further evaluation and determined ineligible for award."

According to a NASA RFP from 2021: "The volume shall be specific, detailed, and complete to clearly and fully demonstrate the offeror's understanding of the Mission Suitability subfactor requirements delineated in paragraph (d) below…."

Now the question becomes one of how to demonstrate understanding by presenting aspects of your proposal with appreciable merit or capability requirements to the considerable advantage of the government during a contract performance (the definition of a significant strength in the Department of Defense (DoD) parlance; DoD *Source Selection Procedures,* August 20, 2022). To rise to the level of a significant strength in NASA procurements, industry's "understanding" must "greatly enhance the potential for successful contract performance" (NASA RFP from January 2023).

Let's pause for a second. Too many times when I review proposals for my customers, in the understanding section, I read something like the following: "The Defense Logistics Agency (DLA) at New Cumberland, PA issued a Network Infrastructure Support Services Request for Proposal (RFP) to upgrade/install the network infrastructure of DLA Information Operations at ten (10) mandatory and four (4) optional distribution locations nationwide." Turns out this was a direct "lift" of language from the RFP. It demonstrated nothing in the way of understanding. It shows rudimentary knowledge of the facts, but does not convey the meaning of the facts, or understanding, as defined by Grant Wiggins and Jay McTighe [1].

So how should we approach understanding in our federal proposals? Understanding is all about your federal government customer—their mission, their success factors and critical issues, the current as-is state of their program or project, and the future to-be state of the program or project. Understanding = Customer.

Over the years, I have developed and continue to hone a series of questions for proposal and capture teams to ask themselves when working to demonstrate understanding. What business processes within the federal agency does this PWS area enable and support? What are the technical critical issues and challenges associated with this task currently? How will those issues and challenges change in the future? How might innovation be introduced to this particular task in close collaboration with the government agency? What would these innovations include? To what other task does this specific task relate, and how? How does this specific PWS task in combination with other tasks support the agency's overall mission? What particular federal mandates, policies, and procedures are related to the governance of this PWS area? How would the particular government agency "paint" a picture (see Figure 9.1) of success on this task now and going forward?

Importantly, the words "we understand" or "we recognize" or "we are aware that" should *not* appear in your proposal. Instead, assert exactly what it is that your organization does understand. Illustrate that understanding in meaningful graphics and tables. Move beyond merely restating the requirements found in the solicitation documents in an effort to demonstrate understanding. That is a quick way to a low evaluation score or rating.

Figure 9.1 Artist brushes on display in a shop in New York City. (Photograph © Dr. R. S. Frey.)

9.2 MISSING THE POINT OF UNDERSTANDING

Lack of demonstrated understanding of the customer's operational environment, challenges, strategic framework, goals, objectives, and overall mission is one of the top two nonprice-related reasons why both large and small companies lose federal government competitive proposals. The other reason is lack of a clear approach that conveys exactly how your company will perform the work.

Recently, I was engaged by a global *Fortune* 500 company to perform an end-to-end review of a series of detailed annotated proposals outlines. This was part of their pursuit of a multibillion-dollar Air Force contract in the western United States. These annotated outlines had been crafted and populated by a team of highly competent and dedicated professionals in their respective areas of expertise. What I observed throughout the outlines across the three major sections that I reviewed was a definite lack of understanding. This despite the fact that the DRAFT Section L, Instructions to Offerors, stated explicitly that: "The Technical Approach/Technical Risk Factor focuses on the Offeror's approach to effectively and efficiently accomplish the requirements of the contract, thereby demonstrating understanding of those requirements...."

Another example of understanding from Section L comes from a 2023 NASA enterprise-wide solicitation: "Demonstrate an understanding of the overall and specific requirements of the proposed Contract." For the video oral presentation, Section L stipulated that the offeror shall provide a discussion that: "Demonstrates a thorough understanding of the technology" and "Demonstrate an understanding of delivering software development using fixed capacity teams." Section M, Evaluation Factors for Award, documented that the video presentation "will be evaluated for overall demonstrated comprehensive understanding...." Clearly, understanding constitutes a critical requirement and dimension of evaluation for the government.

Frequently, proposal authors short-circuit the full intention of the government's request for understanding by simply repeating what is in the RFP or RFQ and that which was learned at various industry days that the government convened.

Companies often begin their understanding sections with a discussion of what the government requires or needs or must have. For example:

- "The government *requires* the intimate support of an industry partner with the knowledge, experience, and resources to help meet the complex, diverse, evolving challenges of a 'no-fail' strategic deterrence mission."
- "Program management for this contract *must* reflect a closely aligned collaboration between government and contractor leadership to ensure sustained superior execution."

Neither statement conveys meaningful, below-the-surface understanding.

Together, we will explore a structured, repeatable process that I have developed to ensure that we convey genuine understanding in our proposals and increase the probability of winning. Specifically, consider asking and answering the following pivotal questions:

Q1: What business processes within the government agency does this task or functional area enable and support?

A case in point: The U.S. Customs and Border Protection's (CBP) Office of Information and Technology issued a solicitation for enterprise data center support services. The data center was located on the United States-Mexico border in California. On the surface, this procurement seemed to be all about IT. However, there were much deeper business and financial dimensions. Turns out that this data center was used to capture critical information about truck traffic between the two countries. Data was used to support the collection of tariffs and taxes. Funds were transferred directly into the U.S. Treasury.

The narrative below is a subset of one of the understanding sections of this proposal. It represents an excellent example of what an understanding section should look like.

> CBP is already facing the need to manage *petabytes* of data. In addition, there is an increasing requirement for storing data that will be used for business and mission support throughout CBP, as well as the Department of Homeland Security (DHS) and other Federal agencies. CBP's Office of Finance's accounting system sits on top of the database within the solution stack, and is critical to the collection of $30 billion in tariffs and taxes each year. CBP has made major investments in Oracle's Exadata database appliance, and already moved approximately 10 major applications to Exadata. Consolidating Oracle

databases into Exadata reduces costs as well as drives down administrative burdens.

Here is a lead-in paragraph that conveyed genuine understanding associated with Task 7, which was one of more than 10 tasks:

> Task 7 supports a critical group of Customer Information Control System (CICS)-based legacy systems that provide the foundation for some of CBP's most mission-essential financial systems. In particular, major financial transactions systems within the CBP Office of Finance are supported in this mainframe environment. Task 7 also involves support for the mission-critical systems involved in the enforcement of U.S. trade and tariff laws. While it is sometimes challenging to integrate these mainframe systems with newer systems and technologies, these CICS-based applications provide CBP with functional, reliable, low-risk systems for some of the Agency's most important activities. In essence, the systems that Task 7 support are essential to all of CBP's revenue collection initiatives, as well as to the national security of the United States.

Q2: What particular federal mandates, policies, and procedures are related to the governance of this task?

Let's look at NASA for one example. At an overarching level, the PMA defines government-wide management priorities for all federal agencies to improve how government operates and performs. In addition, the Program Management Improvement Accountability Act (PMIAA), Public Law No. 114-264, aims to improve program and project management (P/PM) practices within the federal government. A critical governance document for this agency is the NASA Policy Directive (NPD) 1001.0D, the 2022 NASA *Strategic Plan.*

To drill down further with a focus specifically on safety, which is paramount within this agency, there are also policy directives and procedural requirements that govern activities and actions. In addition, there are other federal standards related to safety.

- NPD 8700.1F, NASA Policy for Safety and Mission Success;
- NASA Procedural Requirement (NPR) NPR 8715.3D, NASA General Safety Program Requirements;

- NPR 8715.1B, NASA Safety and Health Programs;
- 29 Code of Federal Regulations (CFR) 1910, "Occupational Safety and Health Standards."

Q3: Who are the stakeholders of this particular customer organization?

Another meaningful way in which to convey understanding without using the expression, "we understand," is to construct a stakeholder diagram. This construct is part of the PMBOK. In this example for the Federal Deposit Insurance Corporation (FDIC) failed bank system, I built this four-part diagram that captured governance and drivers, IT interfaces, stakeholders, and users of failed bank data. Technical SMEs then vetted this for accuracy and completeness. The notable outcome was that a significant amount of insight (understanding) was conveyed in a concise graphic that did not consume more than one-half of a page (Figure 9.2).

Take understanding very seriously. On the internet, there is a wealth of meaningful, government-approved information—agency strategic plans, technology taxonomies, management plans, and learning agendas for all Cabinet-level agencies. Mind these documents to construct on-target understanding sections in your proposal.

Your company's goal is to have understanding rise to the level of a significant strength in the government's source selection findings.

Here are several direct examples:

- "Significant strength for demonstrating a complete and highly effective understanding of the technologies involved in identifying and responding to cybersecurity incidents."
- "The proposal contained a strength for complete understanding of how to address changing requirements, propose test enhancements and implement government-approved enhancements."
- "Significant strength: Presented an exceptionally clear, effective and thorough strategy for cloud migration that demonstrated a significant depth of understanding. This includes outstanding understanding of current cloud computing services, their uses and challenges, as well as a superior plan for cloud migration workforce development through mentoring and supporting internal programs and key staffing."

Figure 9.2 A well-constructed stakeholder diagram demonstrates below-the-surface understanding.

Take note of the government citing that the prime contractor who proposed the cloud computing services discussed the uses and challenges associated with current cloud services. Presenting insight into the challenges is another great way to demonstrate in-depth understanding.

A sizable number of senior managers—particularly in large corporations—convey through their words and their body language the position that...*yeah, yeah, yeah...* we know all about this or that technique or process or approach. That is precisely why I relish the opportunity to work with small organizations to compete against large corporations. Being open, approaching activities with what Zen Buddhism calls the "Beginner's Mind," is essential to increased business and personal maturity. I see evidence of lack of openness when it comes to understanding sections in proposals. You are most effective with an open mind.

I will close this chapter with a compelling understanding section.

NASA today is very different from the NASA of the 1960s. Not only has NASA delivered crucial technologies for society, such as water filtration systems and satellite-based search-and-rescue, it has also evolved its dominant logic and business model. NASA has moved from being a hierarchical, closed system that develops its technologies internally to an open network that embraces global innovation, agility, and collaboration. As the leading agency in cutting-edge scientific discovery and innovation, NASA faces diverse and evolving challenges and opportunities for application, security, and operations that support the agency's missions.

Reference

[1] Wiggins, G., and J. McTighe, *Understanding by Design,* expanded 2nd ed., Upper Saddle River, NJ: Pearson/Merrill/Prentice Hall, 2006.

10

ORCHESTRATING TEAM INTERACTION AND DEVELOPING AN ON-TARGET PROPOSAL APPROACH

Together, people, processes, and tools contribute to the development of a compelling proposal approach. Here, we will discuss the proposal directive (tool), the proposal manager (people), structured interviewing of SMEs (process), rules of engagement for effective meetings (process), Proposal Readiness Work Products for capturing technical approach sections (tool), and enhanced remote team interaction (process).

10.1 PROPOSAL DIRECTIVE—LIVING ENCYCLOPEDIA OF A WIN

Having all of the essential information about a specific proposal contained in one living electronic document marked "Competition-Sensitive" and safeguarded as such is highly beneficial to everyone on your capture and proposal team. I call this document the "Proposal Directive." It is a tool that I conceptualized 23 years ago that still pays dividends for organizations today. I have seen the proposal directive go through 25 iterations on a billion-dollar Department of Energy proposal effort for which I served as proposal manager, so I really do mean that it is living. Please note that the proposal directive is not the same as the actual proposal. Rather, it is a guide to building the proposal volumes. Microsoft SharePoint or Google Drive are two platforms on which all of the proposal volumes are housed securely.

Right on the cover page, the mission of the particular federal government agency stands prominently in the center. I did this to have all proposal contributors keep the customer's mission uppermost in their minds. The names of the capture manager and proposal manager also appear on the cover page for easy reference.

The next pages list the names, contact information (email and mobile phone), time zone where they are located, and specific role of the contributors to the proposal effort. These are followed by a program overview that includes such line items as solicitation number and expected final release date for the RFP, period of performance, total contract value (TCV), and evaluation criteria. Next comes a series of textboxes in which customer hopes, fears, biases, success criteria, and critical issues are presented. A listing follows of customer staff, whether they will be part of the Source Evaluation Board (SEB), and their specific likes and dislikes relevant to the program being pursued.

Next, each team member is listed along with brief corporate profiles (not copied-and-pasted from the company's website) and relevant contractual experience. This is followed by the results of a SWOT analysis of the prime contractor, as well as a competitive analysis diagram that maps knowledge of the competitor to the level of competitive threat. A current government customer organizational chart is essential, as well as populated call or trip reports.

Of critical importance is a listing of validated strengths that your team offers the government customer that are aligned with the Evaluation Factors for Award (Section M). The concise, 2 to 3-page Win Strategy white paper in narrative format comes next. Subsequent to that is your organizational chart for this program, followed by program infrastructure requirements such as facilities and special equipment.

Proposal volume leads, proposal outline, and proposal action items come next. Be sure to give each action item a unique identifier to facilitate clear communication during status meetings. Include proposal production/documentation requirements and proposal deliverables (e.g., platform demonstration and proof of concept), along with a calendar-style proposal schedule. I have found this to be much more accessible and understandable than a Gantt chart schedule done in Microsoft Project. A list of relevant acronyms provides all proposal contributors with a common understanding. Finally, a proposal-specific writing style sheet (e.g., Team ABC versus The ABC Team) and key terms from the government's proposal lexicon (e.g., effectiveness, level of confidence) are valuable elements in your proposal directive.

Develop a proposal directive for every proposal that your company primes (see Figure 10.1). You will be glad that you did.

10.2 THE CRITICAL ROLE OF THE PROPOSAL MANAGER

Over the years, I have assisted organizations to develop outcome-oriented position descriptions for key staff in business development, capture management, and proposal development. Below are excerpts from the most recent one for proposal management.

The proposal management function drives the solutions developed during the capture phase into the proposal volumes. Importantly, the proposal manager "reports" to the capture manager for a given proposal opportunity.

Figure 10.1 The Library of Celsus (Greek: *Βιβλιοθηκη του Κελσου*) is an ancient Roman building in Ephesus, Anatolia, now part of Selçuk, Turkey. (Photograph © Dr. R. S. Frey.)

1. Responsible for managing the set of proposal documents and ensuring exacting compliance with all formatting, structural, and due-date requirements.
2. Designs and builds the proposal volumes to create positive difference between Company ABC and the competitors.
3. Develops the outline and proposal calendar in close coordination with the capture manager.
4. Participates in proposal kickoff and review meetings.
5. Interviews SMEs and management staff to extract and then document key information relevant to a given proposal effort. Insofar as a proposal manager interviewing SMEs, I personally have done exactly that for a small 8(a) company where I worked full time. I trained my other proposal managers to conduct meaningful interviews as well.
6. Interacts and coordinates with teammates, outside consultants (e.g., proposal writers), and vendors (e.g., printing companies) during the proposal development process.
7. In close coordination with the capture manager, sets up "rolling" and/or formal reviews for the proposal, and communicates timelines and specific requirements.
8. Oversees graphics, desktop publishing, and printing of all versions (interim and final) of the proposal volumes.
9. Serves as a key quality checkpoint for acceptance of all text and graphics from internal Company ABC staff, as well as teammates.
10. Remains current on all amendments and modifications to the solicitation and communicates these changes quickly to the entire capture team.
11. Coordinates and tracks data calls across the entire team.
12. Maintains exacting configuration control of all proposal documents and other work products (e.g., graphics and past performance questionnaires) using knowledge-sharing portal.
13. Communicates requirements and timelines regularly with the BD lead, capture manager, volume leads, or book bosses, as well as with infrastructure—employs a "no surprises" approach. Visits capture and proposal staff F2F whenever pos-

sible (or virtually) to ensure effective understanding of expectations, goals, and timelines.

- Goal #1: Supports Company ABC's annual corporate target of booking $YY million in business (see Figure 10.2). Leads and manages end-to-end proposal efforts for one major-impact (>$ZZM) and smaller pursuit opportunities simultaneously, as required.
- Goal #2: Supports the achievement of an overall corporate proposal win rate of >60%, and a competitive range determination threshold of 90% or greater. Actually, insofar as this goal, one company at which I worked for 9 years had a validated proposal win rate of 67%. That was substantiated by banks and attorneys during the due diligence process prior to sale of the firm. I can attest to the rigor with which that win rate was validated. Everything had to be documented.
- Goal #3: The correct version of all proposals as well as interim responses to the government (e.g., requests for information

Figure 10.2 Thoroughbred horses compete at Charles Town Races in West Virginia. (Photograph © Dr. R. S. Frey.)

(RFIs), sources sought) must be delivered 100% on time, every time.

10.3 PULLING INTRIGUING THREADS—INTERVIEWS WITH TECHNICAL SMES

As a small business entrepreneur, your most precious, and limited, resource is time. Your operations staff on the ground have a full-time job keeping their government customers happy. How do you capture and codify the breadth and depth of what your technical staff do and have done while working for your firm? "Street" résumés—the ones that led you and your management team to hire the technical SMEs in the first place—only look backward in time. Let me share a pathway that several small businesses that I support have followed. I know of this because I personally spearheaded this initiative in several firms, one of which had been in business for 10 years but whose leadership was not fully aware of the talent they had working for them nor the span of services they might offer to the government.

These initiatives began with planning between myself and the executive leadership of the small organization. This was followed by leadership's communication with their technical SMEs to inform them about the process and the goals of the initiative. The intent was to generate interest and excitement among the staff. Next, I developed a standard battery of interview questions to ask each SME, either F2F or virtually. Among the 21 questions were the following:

1. Characterize your core competencies that you apply for Company ABC.
2. Tell me about a recent federal government project you worked on as an employee of Company ABC. What were your specific responsibilities and contributions to project success?
3. What automated technical and/or programmatic tools do you use routinely (including version/release numbers) in doing your work? How do they assist you?
4. Provide an example of how you worked around a specific technical challenge that you faced on your government project(s). What did you do, and what were the results?

5. What innovative programmatic and technical approaches, tools, and techniques have you and your team applied to assist your government customers? What have been the results in terms of enhanced quality, schedule adherence or compression, cost control or avoidance, risk mitigation, safety for all staff, and enhanced security profile?

The companies set up dedicated blocks of time for me to interview their technical staff. Interviews were audio-recorded with permission. My focus was not on taking notes, but rather on keeping the conversation going and "pulling intriguing threads" (see Figure 10.3) that emerged during the discussion. Each interview lasted about 1 hour. SMEs had been sent my battery of questions in advance, and most came well-prepared. Interviewees were encouraged to provide

Figure 10.3 Multicolor threads used for sewing. (Photograph © Dr. R. S. Frey.)

me with copies or photos of any commendations (e.g., emails, letters, awards, plaques) that they received from their government customers while with their current employer (my customer). In addition, I was interested in receiving photos of them doing good work for their current government customer, delivering a presentation or conducting training, or receiving an award.

Next, I worked with a third-party service provider to transcribe all of the MP3 audio files into Microsoft Word documents. Each Word file was notated as to who was speaking and line numbering was added to simplify information retrieval. Once the interviews were transcribed and all ancillary materials (e.g., photos and commendations) collected and assembled, I then focused on building individual résumés, formatted in accordance with the most stringent key personnel template the company had encountered among its federal customers. As a baseline, I leveraged the existing corporate résumé for each SME, and then fitted together the newly collected information, photos, and commendations. Each recently crafted résumé was sent back to the individual and their manager for quality validation, and edits were incorporated.

As a follow-on step, I mined the new résumés from those SMEs who, together, were supporting a given project or task in order to build compelling and up-to-date past performance citations. I also built and illustrated concise modules that focused on the company's core technical competencies. These, too, were QC'ed by senior technical managers.

Finally, I reviewed the aggregate of the résumés, past performance citations, and competency modules with an eye toward identifying new potential lines of business for my customers. Turns out, there were.

So, what were the specific benefits of these initiatives? For the first time in each company's history, there were current résumés for each technical SME and updated past performance citations. Both of these sets of documents were used to support fact-based bid/no-bid decisions. They also served as an excellent starting point for résumés and past performance citations in the company's proposals. The technical competency modules supported what I term "Rapid Proposal Prototyping." Instead of issuing internal data calls for each and every proposal the company wanted to pursue, the modules could be assembled to align with the PWS or SOW. Busy operations staff were most appreciative. I know for a fact that these initiatives facilitated much greater proposal throughput per unit time, and the wins followed.

10.4 RULES OF ENGAGEMENT FOR EFFECTIVE PROPOSAL MEETINGS

Each time that I lead or facilitate a proposal meeting, I write the rules of engagement on a white board or easel paper (or share my screen in the case of a virtual videoconference). I then communicate these at the start of the meeting. Depending upon the composition and known propensities of the participants, I also invite everyone in the room to come up front and initial the rules of engagement, signifying their personal commitment to adhering to them. A similar gesture of commitment can be accomplished through the Microsoft Whiteboard digital canvas.

Let's look at the rules of engagement more closely:

1. Mutual respect. No matter their level in the organization or team, every participant will be afforded personal and professional respect. All communication will be in a civil manner. Words, tone, and body language will be appropriate.
2. Every voice will be heard.
3. Focus on sharing ideas—let ideas soar. There is no judgment about right or wrong, good or bad. The goal is winning the proposal. Give multiple ideas the light of day for consideration by the team.
4. No talking over each other. It is important that everyone be attentive to the give and take of the open discussion. Sidebar conversations or two dominant individuals seeking to overtake each other with volume are major distractions to each participant who is not speaking. Active listening in receive mode translates into measurable advancement of concepts and ideas.
5. Silence all smartphones and other mobile devices.
6. The facilitator controls the cadence and the timing of the meeting. I am an enthusiastic practitioner of the Agile approach of "timeboxing." Outcomes are articulated clearly at

> the beginning of the meeting. Those might encompass the identification and validation of candidate strengths for the team, documenting high-probability choices for past performance references to be used in the given proposal, or the development of a high-level organizational structure that shows interfaces with the government customer's organization. Within the allotted timeframe, we concentrate on achieving something complete and meaningful and then evaluate the progress that we have made and identify next steps.

Far too many proposal meetings are rudderless. That would be a good thing to change.

10.5 BUILDING YOUR TECHNICAL APPROACH SECTIONS FOR A U.S. GOVERNMENT PROPOSAL

When I support my customers in thinking about and preparing their technical approach sections, I apply my copyrighted Proposal Readiness Work Product (see Figure 10.4) to ensure a consistent look and feel for government evaluators. Among the critical components in this work product are an elevator speech and a quadrant diagram. Think of the elevator speech as 6 to 8 concise, cogent sentences that highlight the essence of your company's technical offer for the particular task or functional area. In the elevator speech, connect your technical approach to the government agency's mission and operational/business environment. Indicate clearly the value or merit that your technical approach will provide to the agency in terms that the government cares about, and which are evaluated in accordance with Section M (Evaluation Factors for Award) of the RFP. Your company should gain insights into what the government cares about for the specific procurement being pursued through F2F business development interactions. Importantly, because support services contracts are dynamic, that is, they change throughout the period of performance (PoP), ensure that your company's technical approach conveys how you will adapt to shifts in mission, platform, interfaces, service level agreement (SLA) elements, key performance indicators (KPIs), and intended business outcomes driven, for example, by new Office of Management and Budget (OMB) Memoranda.

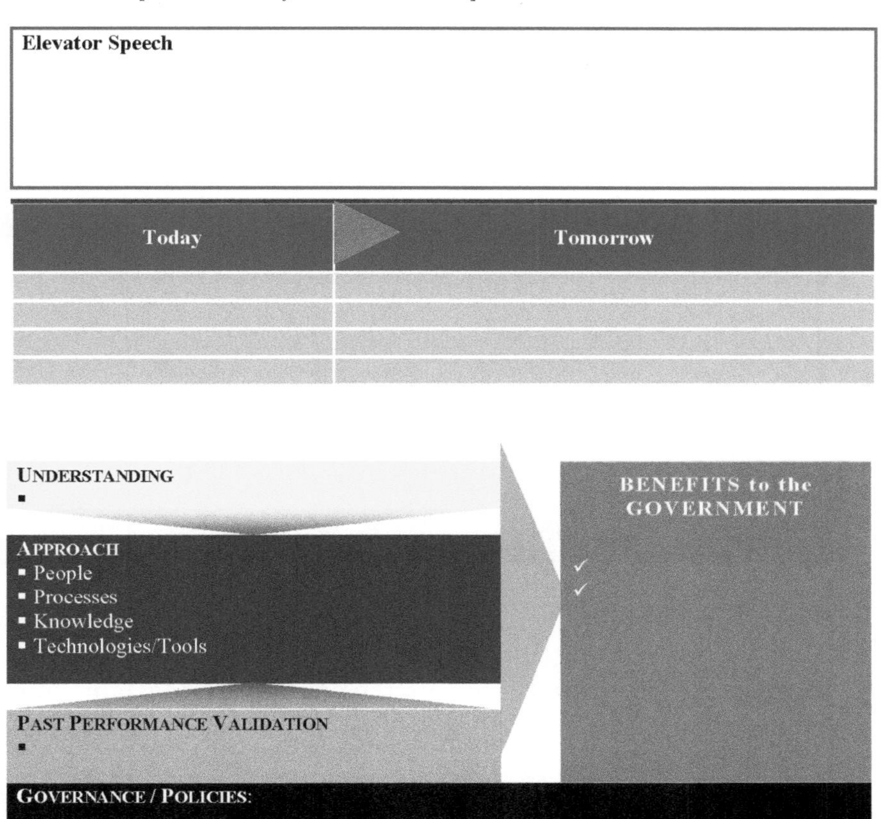

Figure 10.4 Proposal Readiness Work Product. (Conceptualized by Dr. R. S. Frey, and copyrighted through the U.S. Library of Congress.)

The quadrant diagram for the technical approach consists of four segments related to exactly how your company will perform the work for the government agency: (1) people/resources; (2) processes, procedures, and methodologies; (3) knowledge/data; and (4) tools, techniques, and technologies. Too often, I observe companies not focusing sufficient attention on the people who will be performing the work. Therefore, I read such statements as: "Our plan ensures....", when actually, plans, in and of themselves, do not ensure anything. Rather, it is your company's staff professionals who follow the plan

Candidate Innovations	Resulting Efficiencies for the Government

Potential Risks	Mitigation Strategies

Evidence of STRENGTHS
▪ ▪

Figure 10.4 *(continued)*

to deliver repeatable results who are the ones delivering value to the government customer.

As an example:

- *People:* Policy experts, software developers, software testers;
- *Processes:* Hybrid agile project management (APM) framework, automated end-to-end software testing, just-in-time delivery;
- *Knowledge and data:* Healthcare standards, including the Fast Healthcare Interoperability Resources (FHIR) standard; data warehouse from which to generate analytical reports used in decision-making;
- *Tools, techniques, technologies:* Java programming language, CucumberStudio.

Note that approach is not the same as past performance. Yet, I have seen numerous times where companies, particularly incumbents, indicate that they will apply the same proven processes and techniques that have been used successfully during the past 5 years on the new recompetition. Past performance can be leveraged to validate an approach, but it should absolutely not be substituted for a strong characterization of exactly how your company will accomplish the work for the specific customer on the particular procurement at hand.

10.6 INCREASING EFFECTIVENESS IN REMOTE PROPOSAL DEVELOPMENT ENVIRONMENTS

You have most likely heard of the proposal term, “walking the walls.” In today’s work environment, that likely no longer means physically walking along the walls of a large, secure conference room, looking at various proposal artifacts, such as graphics, annotated outlines, or pages of Red Team-ready integrated text and graphics.

Effective remote work collaboration stands as a critical success factor in the proposal environment of 2023. The technology platforms are plentiful. The most participant-friendly ones that I have encountered are Microsoft Teams, Zoom, and Cisco Webex. Each has a Whiteboard feature that enables real-time idea-sharing by multiple people on the conference. Indeed, there are other platforms, such as Verizon BlueJeans and GoTo’s GoToMeeting.

Of note is that what was forced by the COVID-19 pandemic has actually been underway for at least 9 years. According to a 2017 study called the “Human Face of Remote Working,” which Polycom conducted in collaboration with Future Workplace, two-thirds of today’s global workforce takes advantage of the anywhere working model. This is a significant shift upward from 14% in May 2012.[1]

Below are some lessons learned and best practices that I have gained from intensive remote collaboration in support of my customers since March 2020:

- Definitely have a facilitator engaged to keep the conference on track from a time and agenda perspective.
- Proposal tag-up meetings should be at a mutually agreeable time and definitely not 6 or 7 days a week. Take divergent time zones into consideration.

1. This study is from: Wilkie, D., Society for Human Resource Management, 2017, https://www.shrm.org/resourcesandtools/hr-topics/employee-relations/pages/remote-workers-feel-guilty-.aspx.

- Limit required participation to those people who will benefit from and provide meaningful input to the conference. Revisit the number of participants as the proposal response life cycle continues forward.
- Limit the duration of the meeting to no more than 2 hours, with a break built in every hour. Do not go down the road of 8-hour conferences. Deloitte's March 2020 publication entitled "Remote Collaboration—Facing the Challenges of COVID-19," reported that screen fatigue causes the attention span to shorten. Staring at a laptop or desktop screen is more exhausting than F2F interaction.
- Participants should log in 5 minutes prior to the start of the meeting to optimize everyone's time. In addition, active listening is important for each person in order to maximize their contribution.
- Distribute an agenda with clearly defined outcomes at least 12 hours before the meeting. Again, be sensitive to time zone differences.
- Participants must come prepared by reviewing materials in advance and having their action items completed in accordance with the proposal development schedule. Establish this as a key rule of successful engagement.
- Although reading proposal sections aloud is an excellent technique to identify redundancy and ensure that the narrative is clear, in a videoconference limit the number of participants conducting the readthrough to two or three. Focus on the elevator speeches or introductory paragraphs, then the callouts or textboxes, and then the figure captions and table legends.
- Humanize the conferences—time permitting, it is really good to engage participants in a bit of personal discussion about their family, pets, hobbies, and friends.
- While recognizing bandwidth constraints, whenever possible have all participants connect with both video and audio. More than half of communication is nonverbal and body language—having just a circle with a participant's initials, phone number, or low-resolution photo on the screen does not help people to connect fully.

11

USING SECTION M AS A WINDOW

11.1 FEDERAL PROPOSAL DEVELOPMENT—LOOKING THROUGH THE WINDOW OF SECTION M

Let's start by building some context. Part 15.204-1 of the FAR is called the uniform contract format (UCF). Table 15-1 within the UCF defines the A to M framework for competitive acquisitions for federal agencies that follow the FAR, which includes most executive branch organizations. Federal government organizations that do not follow the FAR for acquisitions include the United States Postal Service and Federal Aviation Administration.

Section M, Evaluation Factors for Award, stands as a pivotal element within the UCF. Whenever I evaluate an RFP or RFQ, I begin with a detailed review of Section M. By the way, a subset of those government organizations that do not follow the UCF still do use the nomenclature of "Section M." A case in point is the FDIC. That organization's document, *Understanding the Government Solicitation Bid Package,* indicates that Section M guides the review and assessment of the mission capability of each proposal against the stated factors and subfactors. Evaluation criteria within FDIC solicitations generally include:

- Key personnel;

- Organizational structure and management approach;
- Technical management approach;
- Relevant experience;
- Past performance;
- Transition plan.

Look carefully at the relevant experience and past performance bullet points. They seem to be similar. However, the Government Accountability Office (GAO) does not interpret things that way. Former attorney Richard D. Lieberman in 2021 [1] noted that: "The experience factor focuses on the degree to which an offeror has actually performed similar work. The past performance factor focuses on the quality of work performed."

The U.S. State Department's *Technical Evaluation Criteria and Plan* (April 19, 2019) [2] stipulated that: "The past performance evaluation factor assesses the degree of confidence the Government has in an offeror's ability to supply products and services that meet users' needs, based on a demonstrated record of performance." It continued: "The past performance evaluation considers each offeror's demonstrated recent and relevant record of performance in supplying products and services that meet the contract's requirements." Finally, it stated: "One performance confidence assessment rating is assigned for each offeror after evaluating the offeror's recent past performance, focusing on performance that is relevant to the contract requirements. This is different than just the offeror's relevant experience, which can be established and confirmed objectively."

Importantly, Section M provides a window (see Figure 11.1) into how a given agency will assess industry's proposals. For example, in 2022, NASA competed a 5-year services contract worth more than $500 million. Included within Section M was language that was of direct benefit to offerors:

- B.4 – Program Manager: The government will evaluate the Offerors management plan pertaining to the Program Manager ... for completeness, reasonableness, and effectiveness.
- B.5 – Phase-in Plan: The government will evaluate the offerors phase-in plan for reasonableness, effectiveness, and efficiency.

Figure 11.1 Ornate window with reflections. (Photograph © Dr. R. S. Frey.)

These four keywords—completeness, reasonable, effectiveness, and efficiency—appeared throughout all of Section M. They should have served as a critical call to action for capture and proposal teams; however, I suspect that they did not in most cases.

One of the things that I did was to actually define each of these Section M words in detail. These definitions helped with thinking, writing, and meaningful reviewing.

- *Completeness:* Approach contains all of the steps necessary to support the defined goals; approach addresses all pertinent elements of the RFP; and approach is plausible and logical.
- *Reasonableness:* Approach (people, processes, knowledge, tools/technology) fits within the center's operational environment and cadence.
- *Effectiveness:* Approach will produce the desired objectives; approach will produce positive results in terms of quality, schedule adherence, cost control, and risk mitigation, "doing the right thing."

- *Efficiency:* Approach produces meaningful results per "unit" of resources expended (people, time, equipment), "doing the thing right."

Well-constructed tables and graphics can also carry the message. These elements can succinctly convey how the solutions offered in the management and phase-in sections of the mission suitability volume were indeed "complete," "reasonable," "effective," and "efficient." Doing that helps the government evaluators to quickly discern the value that an organization is bringing. Aligning your proposal with the exact vocabulary of Section M is a critical success factor in the federal marketspace.

Another example of Section M keywords comes from a 2023 U.S. Agency for International Development (USAID) RFP. For FACTOR 1—TECHNICAL APPROACH: PERFORMANCE WORK STATEMENT, we learn that: "USAID will evaluate the feasibility, creativity, and innovation of the Offeror's proposed technical approach as described in the PWS in accomplishing the Statement of Objectives." So the three keywords in Section M are "feasibility," "creativity," and "innovation."

- *Feasibility:* Working definition—Company ABC's approach on Program XYZ is doable in terms of the following parameters: supports the five goals articulated in the USAID *Joint Strategic Plan (JSP) FY 2022 – 2026* (March 2022), adequately integrates with USAID's investments and initiatives in the specific region, level of resources required, level of government oversight needed, and potential risks (e.g., quality, schedule, cost).
- *Creativity:* Working definition—"the tendency to generate or recognize ideas, alternatives, or possibilities that may be useful in solving problems" or communicating with others [3].
- *Innovation:* Working definition—"the application of new ideas to the products, processes, or other aspects of the activities of a firm that lead to increased 'value'" [4]. The U.S. Department of Transportation [5] provides the following additional insight: "accelerating the implementation and delivery of new technologies."

In a high-profile Air Force solicitation released in November 2020, the following verbiage appeared: "Through the Subfactors of the Technical Capability Factor, the Government will assess the Offeror's familiarity, corporate experience, and technical expertise in understanding and successfully performing selected services of the ISC 2.0." "Familiarity," "corporate experience," and "technical expertise" constitute three important Section M words.

- *Familiarity:* The state of having knowledge about something [6]. For German philosopher Martin Heidegger, this word, which comes from the same root as "family," encompasses the ideas of involvement and understanding [7].
- *Corporate experience:* The extent to which the contractor's previous experience with projects of a similar size, scope, and complexity demonstrates its capability to successfully perform the requirements of this task order [8].
- *Technical Expertise:* The abilities to deal with ambiguities, embrace possibilities, and solve extremely complex issues. Technical experts add significant value to a project through their knowledge and substantial experience. As they dealt with a number of similar cases in the past, they developed an ability to help teams solve technical issues. The U.S. Army Corps of Engineers [9] has contributed the following to the definition: "unique or exceptional technical capability in a specialized subject area that is critical to" government customers.

In November 2020, a Marine Corps Recruiting Command (MCRC) Service-Disabled Veteran-Owned Small Business opportunity included the following language in Section M under Factor 1: Technical Approach and Capability: "Qualifications, capabilities, capacity, and comprehensiveness of the proposed strategies and methods for meeting the government's objectives." Let's focus on comprehensiveness, which is often an elusive word to address in federal proposals.

Comprehensiveness reflects the fact of seeing each part as a function of the whole, of not isolating a particular aspect from its context. This word connotes marked by or showing extensive understanding: comprehensive knowledge. Seeing each part of a process, for example, within a larger context is very analogous to systems thinking. Coined by American systems scientist Barry M. Richmond [10], "systems thinking" is "the art and science of making reliable inferences about behavior by developing an increasingly deep understanding of underlying structure." Richmond emphasized: "that people embracing Systems Thinking position themselves such that they can see both the forest and the trees; one eye on each."

Finally, in an Environmental Protection Agency (EPA) RFQ under Volume 1: Technical Capability, the following statement was included, "The vendor's Technical Capability volume must indicate and demonstrate the merit or excellence of the work to be performed or product to be delivered." Of note is that this statement appeared in SECTION A – INSTRUCTIONS, CONDITIONS, AND NOTICES TO SCHEDULE VENDORS, rather than Section M, Evaluation Factors for Award. Also, it turns out that "merit" is a term that is challenging to generate an operational definition for many in industry.

From the U.S. Department of State, The Acquisition Environment (09-25-2019), merit refers to that which "exceeds specified performance or capability requirements." The Defense Advanced Research Projects Agency's (DARPA) *Guide to Broad Agency Announcements and Research Announcements* (November 2016) [11] also contributes to the definition of merit: "potential contribution and relevance to the DARPA mission."

So, look through the Section M window to guide your proposal response. Build in meaningful content that addresses the government's keywords. By all means, make certain that those exact keywords appear in your proposal response across the narrative and in tables and figures. Even consider highlighting them in the cover art for the proposal volumes.

References

[1] Lieberman, R. D., "Experience vs. Past Performance and What About Multiple Offers?" Public Contracting Institute, 2021, https://publiccontractinginstitute.com/experience-vs-past-performance-and-what-about-multiple-offers/#:~:text=The%20experience%20factor%20focuses%20on,quality%20of%20the%20work20performed.

[2] 14 FAH-2 H-360, https://fam.state.gov/fam/14fah02/14fah020360.html.

[3] Padhi, G., "Creativity—An Analysis," *Journal of Research in Humanities and Social Science,* Vol. 9, No. 5, 2021, pp. 73–75, https://www.questjournals.org/jrhss/papers/vol9-issue5/Ser-2/I09057375.pdf.

[4] http://assets.press.princeton.edu/chapters/s9221.pdf.

[5] https://www.transportation.gov/rural/toolkit/maximizing-award-success-introduction-evaluation-criteria.

[6] https://www.britannica.com/dictionary/familiarity.

[7] Turner, P., "Being-with: A Study of Familiarity," *Interacting with Computers,* Vol. 20, No. 4-5, 2008, pp. 447–454.

[8] General Services Administration (GSA), Evaluation Factors, https://hallways.cap.gsa.gov/system/files/doclib/rfqtaskorderevaluationfactors-1622228300.pdf?file=1&type=node&id=70835&force=.

[9] https://www.hnc.usace.army.mil/Missions/Centers-of-Expertise/.

[10] Arnold, R. D., and J. P. Wade, "A Definition of Systems Thinking: A Systems Approach," *Procedia Computer Science,* Vol. 44, 2015, pp. 669–678.

[11] www.darpa.mil/attachments/DARPAGuideBAARA.pdf.

12

PROPOSAL SOLUTION DEVELOPMENT

12.1 SOLUTION DEVELOPMENT OVERVIEW

Early solution development streamlines the entire proposal process. It saves time, precludes significant rework after formal color review cycles, and reduces overall bid and proposal (B&P) costs—all of which are vital to small businesses in this era of hypercompetitiveness. Your goal should be to begin solution development at the draft RFP stage of the federal acquisition cycle for a specific opportunity. You do not need a final RFP to construct and refine many of the elements in your solution. For example, you can craft your overall corporate-level approach to risk management, including your risk officer or lead staff person, risk planning methodology, and specific quantitative (e.g., composite risk index) and qualitative risk assessment and estimation processes. In addition, critical chain-based project management and risk tracking and analytical tools, along with knowledge-based risk management and risk mitigation and avoidance success stories, are not directly dependent upon the requirements in a government solicitation document. To be sure, there will need to be connections drawn between your overall risk management approach and the potential project-specific risks that may occur during contract execution. However, the essentials of your approach can be established and refined well in advance. The same is valid for your company's total compensation plan (TCP), subcontractor management approach, phase-in/phase-out (transition) plan, and cost and schedule control processes and tools, as well as many other proposal sections, including your plan for recovering from unanticipated problems and corporate training/cross-training programs.

12.2 SOLUTION DEVELOPMENT TOOLS

Other important solution development tools encompass 3-column pain tables, in which are captured customer pain points and other key elements of understanding such as the as-is and to-be operational environment, your approach, and benefits to the government of your approach in terms of enhanced quality, timeliness, cost control and avoidance, and risk mitigation. Innovation tables, risk tables, and elevator speeches, as well as process flow diagrams, features/benefits tables, success stories and proof points, and a listing of validated key strengths that your company brings to the specific customer represent other valuable and field-proven tools through which to capture and illustrate your solutions.

12.3 SOLUTION DEVELOPMENT PROCESS

One of the most effective and efficient ways by which to extract key information that will contribute to the overall solution set of your proposal is to conduct and facilitate a series of highly interactive solution development strategy sessions. These sessions should include representatives from within your own company, as well as from across your entire team.

Participants in these meetings should come fully prepared, having read and studied the solicitation documents in detail. They should be knowledgeable in one or more of the following areas: (1) the customer's as-is and to-be operational environment; (2) the customer's hopes, fears, biases, critical issues, and success factors; (3) project-specific technologies and tools, along with particular technical and programmatic processes; (4) major business and governance drivers, including statutes, regulations, directives, policies, and standards for this specific government customer; and (5) relevant corporate past performance and key resources such as centers of excellence and corporate innovation centers that can be applied directly to the specific contract. Additionally, people who attend these solution-generating sessions must be fully engaged and interactive.

One key rule of engagement is no mobile phones, smart watches, or other personal digital assistants (PDA) of any kind. An experienced meeting facilitator is critical to obtaining meaningful and substantive outcomes that can then be used to develop sections of the proposal effectively. That individual must be able to draw ideas and information out of the heads of a broad spectrum of technical SMEs, executive leaders, project managers, and business development staff. On one Department of Homeland Security (DHS) proposal effort that I supported, we collected vital information through the efforts of approximately 15 people during the course of 3 days of intensive, F2F interaction. Every one of the solution sets for the 13 tasks was addressed and documented in a consistent manner using the same categories, or "buckets," of information as a standard framework. This framework was built from Sections, C, H, L, and M of the solicitation document, as well as the PPQ. The PPQ is always a rich source of information as to the elements the government deems important, such as quality of service, timeliness, efficiency and effectiveness, key personnel management, and cost control, along with minimizing government oversight of the project.

This information-collection practice greatly facilitates the next step in the proposal development process, namely, the actual writing of the elevator speeches, generating process flow diagrams as real artwork, refining features/benefits tables, and crafting full-scale narrative for the understanding and approach sections of your proposal. The more relevant big-picture insights as well as quantitative and qualitative details that you capture during the solution development process, the more powerful and highly rated by the government evaluators that your proposal sections are likely to be. Importantly, technical details alone are insufficient to win proposals. Make certain that the solution development strategy sessions focus significant attention on the relationship across technical task areas. Is one or more tasks or functional areas subordinate to another? Are several tasks the logical extension of another task? Are certain tasks—such as those relating to project management, quality control, and configuration management—cross-cutting, in that they relate across multiple other task areas?

As you conduct your solution development strategy sessions, it is necessary that participants fully grasp the key differences as well as the connections between understanding and approach, the topic of Chapters 9 and 10.

12.4 FRAMEWORK FOR TECHNICAL SOLUTION DEVELOPMENT

Solutions are the essence of your final, written proposal or oral presentation to the federal government for competitive opportunities. A winning proposal requires carefully constructed, customer-focused solutions. Every element in a well-crafted proposal involves a choice, and each choice results in a part of your overall solution. The solution set constitutes all of the choices your company makes with regard to its teaming strategy, technical approach, management and subcontractor management approach, key personnel and staffing selections, phase-in/phase out (transition) timeline and processes, quality management methodology, and risk management approach, as well as past performance references and cost/price strategy. Importantly, develop your solution set early in the capture process—prior to the release of the final solicitation documents so that you can validate the key dimensions of your solution with your government customer. Refine your solutions through customer feedback as well as ongoing market research and meaningful insights provided by companies with which you are teamed on the specific marketing opportunity.

Let's zero in on technical solutions. The engineering category of Capability Maturity Model Integration (CMMI) for development (CMMI-DEV) is composed of five process areas (PAs), one of which is technical solution (TS). CMMI-DEV provides a comprehensive, integrated set of guidelines for developing services, as well as products. As your company develops its technical solution set, you might consider each of the key elements of the TS process:

1. Meet requirements/requirements traceability (documented thread that provides forward and backward visibility into all activities surrounding each requirement in the PWS or SOW)—it sounds straightforward, but it is actually quite involved.

2. Current versus new technologies—do you want to leverage the technologies in which the government has invested already, offer "bleeding-edge" tools, or a combination?
3. Employ techniques and methods known for producing effective design (e.g., prototypes, structural models that consist of the objects in the system, and the static relationships that exist between them).
4. Interfaces/compatibility with the government's systems—technical interfaces are always important to government customers.
5. Criteria to select the best alternative among several available—provide the rationale for your selection.
6. Standards and guidelines (e.g., National Institute of Standards and Technology (NIST), CMMI, Information Technology Infrastructure Library (ITIL)).
7. Documentation—provide what, how, and in what format you will codify your approach.
8. Project planning and customer visibility during execution.
9. Measurements and metrics program to provide thorough insights and facilitate factual decision-making processes—how will you measure success?
10. Implementation roadmap—step-by-step articulation of how you will deploy your technical solutions.
11. Potential risks and opportunities associated with deployment of your solution.
12. Performance categories—cost reduction, schedule adherence, productivity increase, superior quality, and ROI.

Applying this CMMI-based checklist will help your company to develop comprehensive technical solutions.

12.5 BOX-IN-A-BOX MODEL FOR PROPOSAL DEVELOPMENT

Yes, I read the following 35-word sentence in the understanding section of a federal contractor's proposal: "The [Government Agency ABC] in Pennsylvania issued a Network Infrastructure Support Services Request for Proposal (RFP) to upgrade/install the network infrastructure of Information Operations at ten (10) mandatory and four (4) optional distribution locations nationwide." First, sentences in proposals should be between 18 and 25 words to optimize readability and retention. Second, and most importantly, this sentence was an exact quote from the RFP in question. It presented nothing in terms of understanding.

When demonstrating understanding in proposals, a helpful technique that I use with my customers is to invite them to think in terms of something I call the "box-in-a-box" model. Let's take a simple example—the placement of the Washington Nationals baseball team in context. The Nationals are part of the National League (NL) East Division, which, in turn, is part of the National League. The National League is one of the two leagues that constitute Major League Baseball (MLB) in the United States and Canada. MLB stands within the larger framework of North American professional sports teams, which include the National Football League (NFL), National Basketball Association (NBA), Women's National Basketball Association (WNBA), National Hockey League (NHL), Major League Soccer (MLS), and National Women's Soccer League (NWSL).

Connecting with the world of design, the Finnish-American architect Gottlieb Eliel Saarinen was known for his work with art nouveau buildings in the early years of the twentieth century. He said to: "Always design a thing by considering it in its next larger context—a chair in a room, a room in a house, a house in an environment, an environment in a city plan."

So, to translate these same lines of thinking to proposals, the U.S. federal government operates within a defined governance framework (see Figure 12.1) composed of such elements as statutes (e.g., Federal Information Security Management Act (FISMA)); executive orders (e.g., Executive Order 14035: Diversity, Equity, Inclusion, and Accessibility in the Federal Workforce); directives and circulars (Office of Management and Budget (OMB) Circular A-130, "Managing Information as a Strategic Resource); policy directives (e.g., NASA Policy Directive (NPD) 5000.2D, Small Business Subcontracting Goals); instructions (e.g., Department of Homeland Security's (DHS) Instruction Number 1022-01-103, Systems Engineering Life Cycle (SELC)); strategic plans (e.g., DHS Strategic Plan for Fiscal Years 2020–2024); and technology roadmaps and taxonomies (e.g., 2020 NASA Technology

Figure 12.1 Iconic Hangar One at NASA Ames Research Center (ARC) in Mountain View, California. (Photograph © Dr. R. S. Frey.)

Taxonomy). To demonstrate your company's understanding within a specific proposal, show how the specific technical requirements in the RFP fit within the broader context of the agency's: (1) operational requirements, (2) capability gaps, (3) business needs, (4) organizational mission needs, (5) strategic goals, (6) annual performance reporting, and (7) stakeholder community.

Then, in your approach sections, indicate in detail how your company will assist the government agency to conform to the governance framework in which it must operate now and going forward.

12.6 SOLUTION DEVELOPMENT—EFFECTIVE ALTERNATIVE TO BRAINSTORMING

Four years ago, I was faced with organizing a group of very accomplished astronomers, senior scientists, engineers, operations experts, test pilots, and managers at NASA Ames Research Center in Mountain View, California. An initial goal was to converge on a concise picture of the scientific value of the Stratospheric Observatory for Infrared Astronomy called SOFIA (see Figure 12.2). This highly modified 747SP aircraft, a joint NASA and German space science effort, flew above 99% of the Earth's atmosphere with sophisticated telescopes and other instruments on board to study infrared astrophysics. Infrared astronomy allows scientists to observe regions of space that are otherwise obscured by cosmic dust.

Due to the mix of personalities and varying levels of willingness to advance individual points of view, I elected to employ the technique of brain-writing rather than brainstorming. "[B]rain-writing provides everyone with equal time to think and write, and it virtually eliminates pressure towards group conformity. [I]t helps participants if previous brainstorming sessions have been monopolized by one or two dominant members" [1].

More than 2 years later, the 5 × 8-inch index cards on which group members had anonymously written their three sentences of input were revisited. It turned out that many of those initial ideas regarding the value of SOFIA were still in the forefront of writing

Figure 12.2 SOFIA, the Stratospheric Observatory for Infrared Astronomy, is a Boeing 747SP aircraft modified to carry a 106-inch reflecting telescope. SOFIA was made possible through a partnership between NASA and the German Space Agency (Deutsches Zentrum für Luft- und Raumfahrt (DLR)). (Photograph © Dr. R. S. Frey.)

and discussion. Individuals conveyed that they liked the brain writing process as well.

This same creativity technique can be applied to solution development during the proposal life cycle. SMEs can certainly have sharply divergent points of view regarding technical solutions. Brainwriting affords each SME the opportunity to document his or her contribution to the technical issue at hand. On the SOFIA program, I collected the handwritten index cards from each individual after 10 minutes and then redistributed them to another member of the group. No one could update their own card. My next instruction was for the new recipients to add three more sentences that built upon and connected with what the original author had written. No one was allowed to simply cross out the original handwritten text and replace it with new thoughts. Everyone had an additional 10 minutes of composition time. Then each revised index card was read aloud by a team member who was not involved in providing either portion of written input.

Brain-writing encouraged concise and focused thinking and writing. It gave "light" and "air" to everyone's point of view without regard for professional hierarchy. As a team, we were then able to triangulate on the most salient and well-articulated outputs. In the proposal world, highly collaborative solutions that involve the construction of ideas on ideas can similarly be developed using brain-writing.

12.7 HELPFUL PROPOSAL PUBLICATIONS

Over the years, I have discovered or been introduced to a number of publications (see Figure 12.3) that I use on a regular basis when preparing proposals and oral presentations.

- Bounds, A., *The Jelly Effect: How to Make Your Communications Stick,* Chicheser, U.K.: Capstone Publishing, 2007. "Audiences don't care what you say. They only care what they are left with AFTER you've said it." "The most interesting, most important, most critical fact to tell a customer...is what they will be left with AFTER you've done your work."
- Letavec, C. J., *The Program Management Office: Establishing, Managing and Growing the Value of a PMO,* Fort Lauderdale, FL: J. Ross Publishing, 2006. Importantly, Mr. Letavec recommends developing a PMO value statement to help frame the

Figure 12.3 Several published books that have proven helpful for solution development. (Photograph © Dr. R. S. Frey.)

objectives of this office in terms of items that can be implemented quickly. Framing the PMO as a knowledge organization is the author's most significant contribution to this topic. The knowledge role of a PMO includes activities "associated with acquiring, organizing, maintaining, and disseminating organizational knowledge."

- Hilger, J., and Z. Wahl, *Making Knowledge Management Clickable: Knowledge Management Systems, Strategy, Design, and Implementation,* New York: Springer, 2022. Hilger and Wahl developed the Knowledge Management (KM) Action Wheel, which includes create, capture, manage, enhance, find, and connect. The create segment of the wheel recognizes that a key element of good KM is the "creation of new knowledge," not only the "capture of existing knowledge." The authors recommend developing a KM Transformation Roadmap in which not all work is centralized, but that "it is coordinated, aligned, and integrated to achieve the KM Target State."

- Augustine, S., R. Cuellar, and A. Scheere, *From PMO to VMO: Managing for Value Delivery,* Oakland, CA: Berrett-Koehler Publishers, 2021. The authors characterized the Agile Value Management Office (VMO) as an "end-to-end, cross-hierarchy team driving business agility." They invite us to understand "organizations as adaptive networks of teams organized around specific goals." "At its core, your VMO should be organized as a fulcrum that translates your organizational strategy into outcomes-focused action and change."
- Bounds, A., *The Snowball Effect: Communications Techniques to Make You Unstoppable,* Chichester, U.K.: Capstone Publishing, 2013. British author and speaker Andy Bounds invites readers to begin their presentations and emails, for example, with the "DO." What is it that you want your audience to do? What action do you want them to take? Bounds suggested saying the why before the what. Leverage the word "you." Recognize that the other person (not you) perceives what you are offering is a benefit, or not. Make sure that the benefit you are offering is related to their future.
- Pink, D. H., *To Sell Is Human: The Surprising Truth About Moving Others,* New York: Riverhead Books, 2012. Essentially, what we in the workforce are doing is "persuading, convincing, and influencing others to give up something they've got in exchange for what we've got." "To sell well is to convince someone else to part with resources—not to deprive that person, but to leave him better off in the end."
- Kritzer, D. M., *How to Play the Federal Contractor Game to Win: The Unwritten Axioms. A Primer for Selling Services to the Federal Government,* Potomac, MD: David M. Kritzer, 2008. David Kritzer is one of the most talented capture managers I know. He observed that one "must constantly lay the groundwork for future business." In addition, Kritzer introduced the

important term "Dot Distance," and why it is important to let the listener (of an oral presentation) or reader (of a proposal) to connect those dots.

- Kritzer, D. M., *More Solutions for Winning the Federal Contractor Game: Practical (and Humorous) Advice from the Frontlines,* David Kritzer: Potomac, MD, 2014. This book is made extremely accessible and engaging through the use of humorous illustrations and pithy quotations. My favorite is the cartoon showing Farmer McGregor and Peter Rabbit. Mr. McGregor is in a field reading the book, *Peter Rabbit,* and Peter is directly underneath in his burrow reading the *Farmer's Almanac.* This cartoon is accompanied by two questions that one should constantly be asking when working on an opportunity.
- *Practice Standard for Project Risk Management,* Project Management Institute, Inc., 2009. This publication addresses risk management as it is applied to single projects only. It provided a project risk management process flow diagram, discussed both qualitative and quantitative risk analysis, and presented critical success factors for the monitor and control risks process.
- *Agile Practice Guide,* Project Management Institute, Inc., 2015. High-uncertainty projects have high rates of change, complexity, and potential risk. Agile approaches were developed to explore feasibility in short cycles and to adapt quickly based on evaluation and feedback. After introducing the Agile Mindset and Manifesto, the volume discusses the shared heritage among Lean, Agile, and Kanban. The role of the project manager as a servant leader in an Agile environment is presented.
- Patel, N., *Practical Project Management for Engineers,* Norwood, MA: Artech House, 2019. Nehal Patel rightfully began her book with a discussion of communications management, given the criticality of effective communication to project (and proposal) success. She then moved through scope management, schedule management, requirements management, risk management, project resource management, and vendor management. Ms. Patel also addressed cost management, configu-

ration management, and quality management, before offering insights gained from real-world situations.

- Pinvidic, B., *The 3-Minute Rule: Say Less and Get More from any Pitch or Presentation.* New York: Portfolio/Penguin, 2019. Mr. Pinvidic advocated using the WHAC filter for presentations. W: What is it? Your core concept and being able to help your audience understand the fundamental elements the way that you do. H: How does it work? A: Are you sure? The facts and figures that validate your offering. C: Can you do it? Importantly, he advised saying less to get more.
- Heath, C., and D. Heath, *Made to Stick: Why Some Ideas Survive and Others Die,* New York: Random House, 2008. Brothers Chip and Dan Heath wrote that "[s]peaking concretely is the only way to ensure that our idea will mean the same thing to everyone in our audience." Their most "sticky" idea to me was contrasting what President John F. Kennedy actually said in 1961 with what he might have said had he been a chief executive officer (CEO) in a large corporation. Kennedy actually called "to put a man on the moon and return him safely by the end of the decade." This call was simple, unexpected, and concrete. If he had been a CEO, President Kennedy might have said: "Our mission is to become the international leader in the space industry through maximum team-centered innovation and strategically targeted aerospace initiatives." What a profound difference!
- *Practice Standard for Work Breakdown Structures, Third Edition,* Newtown Square, PA: Project Management Institute, Inc., 2019. This resource defined a work breakdown structure (WBS), identified four concepts that apply when creating a WBS, and discussed using a WBS in waterfall and Agile life cycles.
- Parkinson, M. T., *Billion Dollar Graphics: 3 Easy Steps to Turn Your Ideas into Persuasive Visuals,* PepperLip Press, 2006. My professional colleague and personal friend, Mike Parkinson, set the stage for his well-illustrated, example-filled book in the following way:

Graphic Description	Textual Description
	A curved line with every point equal distant from the center

Parkinson introduced the P.A.Q.S. mnemonic—primary objective (P), audience (A), questions (Q), and subject matter (S). "Knowing your P.A.Q.S. is vital to the success of your graphic."

Check out these books—each will be a valuable addition to your company's proposal library.

Reference

[1] Litcanu, M., et al., "Brain-Writing vs. Brainstorming Case Study for Power Engineering Education," *Procedia—Social and Behavioral Sciences,* Vol. 191, 2015, p. 390.

13

PRECURSOR TO THE EXECUTIVE SUMMARY

13.1 INTRODUCTION

There are times when the federal government requires an executive summary, even stipulating that it be a stand-alone proposal volume. In other cases, an executive summary is not an explicit requirement. In any event, the important thing is that whatever it is called (e.g., introduction, overview), the executive summary be embedded in a section of your proposal that gets evaluated. That might be at the beginning of mission suitability in a proposal to NASA or within the structure of mission capability in a proposal to the Air Force.

Now, as to when to write the executive summary, there are several schools of thought. My experience suggests that it is best to begin crafting the executive summary in the pre-final RFP stage—in effect, early. Then keep maturing that document throughout the proposal development life cycle, finalizing it after the last major review cycle.

My recommendation is to compose what I call a "Win Strategy white paper," which is the precursor to the executive summary (see Figure 13.1). This is a capture management responsibility, with input from knowledgeable business development, programmatic, technical, and executive staff. Leverage the draft RFP as guidance to build the structure of this Win Strategy white paper, or the proposal or final

Figure 13.1 Cherry wood blank mounted on a wood lathe. Using very sharp, high-speed steel chisels, the craftsperson turns the blank into a finished chalice (i.e., the executive summary). (Photograph © Dr. R. S. Frey.)

RFP from the last competition cycle. These may have to be obtained through an FOIA request. Given today's significant backlog of FOIA requests across the federal government, build in a long lead time for this process. The absence of a draft RFP should not stand as an impediment to crafting the Win Strategy white paper, as discussed below.

One way to think about building the Win Strategy white paper is to draw upon what we know already about most federal support services solicitations. Namely, there is a: (1) technical approach [T], (2) management approach [M], (3) staffing plan [S], (4) phase-in/phase-out plan [Ph], and (5) past performance [P] section (and cost/price, to be sure). There could be a requirement for corporate resources along with other plans, such as Safety, Health, & Environmental (SH&E); Organizational Conflict of Interest (OCI); Diversity, Equity, Inclusion, and Accessibility (DEI&A); Total Compensation Plan (TCP); External Customer; and Technologies, Innovations, and Process Improvements (TIPI). Items (1) to (5) might be framed in agency-specific language and appear in varying sequence in a particular RFP, but the overall intent and focus is the same—T-M-S-Ph-P.

Given this knowledge, the Win Strategy white paper can be constructed well before release of the final RFP. Target developing 2 to 3 pages of concise, cogent narrative, versus bullet points only. This practice helps to ensure that your organization's overarching story for the specific procurement holds together as a clear, cohesive, credible, and compelling whole. As the Win Strategy white paper is matured, quantitative proof points that validate your organization's offer should be embedded in appropriate sections. Leverage the specific Section M language from the final RFP to help the government evaluators "feel at home" in what will become the executive summary. Importantly, season the executive summary with insights gained from business development efforts with the customer decision-makers and influencers.

The great thing is that the Win Strategy white paper can and should serve as a guidepost for content providers/writers, as well as reviewers. It definitely helps to ensure consistent and logically connected messaging.

13.2 WIN STRATEGY WHITE PAPER AS A SOLUTION DEVELOPMENT TOOL

Solution development tools include a Win Strategy white paper, which should document the essence of your offer across all major sections of your proposal. It is the precursor to the executive summary, and it serves as a forcing function for your company and your team to think through and architect solutions in advance of the actual proposal response. We know that, in any given proposal, there are a number of fundamental and predictable sections, even though there may be permutations in the exact titles across various solicitations (e.g., management approach (used in some Department of Energy solicitations), management plan (used in some Department of Justice RFPs), management organization and management systems (used in some Environmental Protection Agency solicitations), and program management approach (used in some intelligence community RFPs)). Importantly, we do not need a draft or final RFP to know this. These sections include:

- *Technical approach:* Understanding the requirements, approach, and tangible and intangible (e.g., attention to detail, peace of mind) benefits to the customer.
- *Management approach:* Including organizational structure, performance and quality management, schedule management, and cost management, as well as resource management, staff development/training, communications plan, knowledge-sharing plan, specific mechanisms to access corporate assets, and risk management.
- *Staffing plan:* Key personnel, recruiting and retention approach, total compensation plan (TCP).
- *Phase-in/phase-out:* Transition approach for both incumbent capture and an alternative to incumbent capture, in effect, a Plan B.
- Subcontractor management, subcontractor interfaces, and subcontractor utilization, along with teaming configuration and rationale for teaming.
- Quality management plan.
- Risk management and mitigation.
- Past/present performance.
- Cost/price and bases of estimates (BOEs).

Structure your Win Strategy white paper to articulate and document your key solutions (choices) for each of these major sections. Collectively, these constitute your solution set. In the case of past performance, for example, identify the highly relevant projects that you will cite that are similar in size, scope, complexity, and contract type, as well as the government staff (e.g., contracting officer, contracting officer's technical representative, and technical monitor) or private-sector managers who will be listed as key points of contact and references for these same projects. For your management approach, discuss whether you will implement an integrated project team (IPT) or an organizational framework that is constructed with specific teammates focused on particular functional or task areas (or a hybrid management configuration). When developed with enough lead time prior to the release of the final solicitation document from the government, key solutions that are captured in your Win Strategy white paper can then be vetted with appropriate government staff and modified and refined accordingly. Making the government informally aware of your solutions prior to the release of the final RFP is accomplished during F2F or virtual business development customer visits.

Your goal needs to be to make the government aware—informally, not in writing—of your solution set across all of the aforementioned sections before they read about it in your proposal. Remember that the proposal is the final exam in the overall business development process—your solutions must be presold. The content of the Win Strategy white paper should be based on many of the key elements of your proposal directive, including SWOTs; call reports; and customer hopes, fears, biases, critical issues, and success factors. This white paper is a narrative that integrates the various components of the directive into a cohesive story to be used by proposal writers and contributors in the event that there is a final go bid decision made (see Figure 13.2). The business development and capture management functions in your company should work on this deliverable collaboratively to ensure that there is effective communication and inclusion of all of

Figure 13.2 Aerial view of the new Point Loma Lighthouse on Coronado Island near San Diego, California gives a big-picture perspective of the area, just as the Win Strategy white paper does of your proposal solutions. (Photograph © Dr. R. S. Frey.).

the details learned during the course of the business development process. The Win Strategy white paper should be approximately 2 to 3 pages in length and can be leveraged by executive management for multiple decision-making purposes. These include adding niche subcontractors or unpriced vendors, final bid/no-bid, and investments in staff, hardware, software applications, facilities, and/or training that may offer significant strengths and thus improve the overall value of your offering to the government.

13.3 THE EXECUTIVE SUMMARY AS A REQUIREMENT

Although it is not common, there are times when U.S. federal government agencies include a specific provision for an executive summary in their RFPs:

- U.S. Trade and Development Agency (USTDA) (January 2023): The RFP instructed that offerors should include an executive summary in their proposal and, within their summaries, include such detail as proposed key personnel.
- U.S. Department of the Army (September 2020): Executive Summary (Volume I): The executive summary shall include a brief summary of the offeror's capability and approach to accomplish the requirements of this contract.
- U.S. Army Corps of Engineers (USACE) (April 2015): The offeror shall provide one executive summary to describe the best example of its work for a detailed audit report completed for government (preferably a Department of Defense activity) or a private sector client.
- U.S. Department of Energy (DOE), National Nuclear Security Administration (NNSA) (August 2021): An Executive Summary or Overview of Volume II may be provided in Volume II and is subject to the 25-page limitation.
- U.S. Agency for International Development (USAID) (January 2023): C. Executive Summary (1 page): The executive summary must provide a high-level overview of key elements of the technical application.
- NASA (March 31, 2022) Broad Agency Announcement (BAA): 4.1.2.2 Executive Summary: Describe the proposal's prominent

> and distinguishing features. The executive summary should provide an overview of the proposed effort that is suitable for release through a publicly accessible archive should the proposal be selected.

In these types of cases, offerors must include a concise executive summary and name it clearly as such. There are always exceptions to the general rules.

14

MULTIPLE PALETTES

14.1 ANOTHER PALETTE: A MEANINGFUL PROPOSAL COVER LETTER

To be sure, there are times when the federal government RFP or RFQ stipulates that only a transmittal letter be provided, and that is it. Simply convey what documents are contained in the electronic upload or the hand-delivered box of hardcopy notebooks.[1] No other information is required or allowed. In other instances, the government defines precisely what is to be included in the proposal cover letter: for example, authorized offeror personnel, company/division street address, company/division's Unique Entity Identifier (UEI)[2] and Commercial and Government Entity (CAGE)[3] code, period for acceptance of offers, and summary of exceptions.

1. Yes, even today, agencies such as the Department of Veterans Affairs still require delivery of hardcopy business and technical volumes. A recent case in point was the Remote Patient Monitoring-Home Telehealth solicitation, proposals for which had to be delivered in person to Golden, Colorado.
2. UEIs are created in the U.S. government's System for Award Management (https://sam.gov). On April 4, 2022, the General Services Administration (GSA) began using the UEI as the authoritative identifier for all entities doing business with the federal government and discontinued using the Data Universal Numbering System (DUNS). This was a proprietary system developed and managed by Dun & Bradstreet.
3. The Defense Logistics Agency (DLA) maintains the CAGE website.

However, when no details are stipulated, and the proposal cover letter stands outside of the core page count, industry has a golden opportunity. Yet, most times what I read is: "Company ABC is pleased to submit our response to RFP 80-23-R0001. Enclosed please find the requisite copies of Volumes I, II, and III. Should you have any questions or concerns, do not hesitate to contact me at…."

When meeting F2F with a senior U.S. Army Contracting Officer in Crystal City, Virginia, she told me very directly that the government does not care at all whether a company "is pleased to submit" anything. Rather, she said, what the government (and in her case, the Army) does care about is how that company will support the mission—namely, the warfighter and combatant commanders.

Let's look at an alternative approach, which still fits within 1 page:

> Expanding today's engineering boundaries to support science and space exploration stands at the heart of the mission for Civilian Agency X. That is precisely where we bring extensive and tangible value to you. Our culture drives investment in technical capabilities and workforce development for this contract. We support continuous improvement and reinvestment in the technical currency of the workforce in whom Civilian Agency X has invested deeply, as well as in technology infusion and innovation.
>
> Past performance volume: As a team, we bring the best practices, lessons learned, and insights gained through our three very highly relevant contracts at X, Y, and Z.
>
> Mission suitability volume: Subfactor A: Company ABC's effective, low-risk approach builds on our extensive experience in spacecraft engineering and space flight hardware systems. Our suite of proven innovations provides effective evaluation of design options, and reduces development times, labor hours, and required resources.
>
> Mission suitability volume: Subfactor B: Our process-driven management approach is informed by the "GOOD PRACTICES" of PMBOK 7th Edition and focuses on delivering highly effective and efficient engineering support. We measure and manage performance per the ISO 9001:2015 standard, our AS9100D quality management system, CMMI best practices, adherence to Civilian Agency X timelines, our record of cost control, and effective risk mitigation. We designed our integrated management approach around maintaining

continuity of current operations during Phase-In, then successfully executing simultaneous Task Orders to achieve project and mission goals.

Whet the government's appetite to evaluate your organization's proposal. Some folks may say that the proposal cover letter does not get read. I suggest that any time industry has a palette (see Figure 14.1) through which to tell its story, take full advantage of it.

14.2 PROPOSAL GRAPHICS AND PROPOSAL COVER DESIGN[4]

As human beings, we take shortcuts to make choices. We do judge a book by its cover. We are influenced by what we see. We are hard-wired to react differently to visuals. Our life experience has proven that what we see and experience (feel) are strong indicators of quality. For example, imagine you walked into a dirty restaurant with the ceiling, floor, and furniture stained and falling apart. Is this a high-quality restaurant? Would you want to eat there?

In the same regard, poorly rendered graphics can negatively influence an evaluator's perception of the solution and, by association, the presenter (the person, place, or thing most associated with the graphic) because of these neurological and evolutionary traits. The reviewer's understanding and opinion of our content along with their emotional state while evaluating our proposal influence the decision they will make. Just like entering a well-maintained and clean restaurant, professional graphics that are clear and concise elicit a positive impression quickly. Using professional and easy-to-follow visuals is one way to engage and motivate the evaluator to choose your solution. Using professional, helpful graphics reassures reviewers that you are competent and capable. It gives you an advantage over your competition who is not investing in the same effort with their visuals.

I have spoken with more than 100 government evaluators, reviewers, and decision-makers. Overwhelmingly, they agree that visuals, when done properly, help them do their job. For example, Jacob Bertram, a federal government contract operations at the General

4. This section was written by Mike Parkinson. He is a partner at the 24 Hour Company (www.24hrco.com), a premier creative services firm, and owns Billion Dollar Graphics (www.BillionDollarGraphics.com).

Figure 14.1 A spectrum of watercolor paints. (Photograph © Dr. R. S. Frey.)

Services Administration (GSA), Federal Acquisition Service (FAS), Defense Contract Management Agency (DCMA), and Defense Contract Audit Agency (DCAA) was asked, "Do evaluators really read what is written, or do they focus on graphics or callout boxes?" Bertram replied with the following:

> [Many] evaluators ... are pressed for time. Proposals are not meant to be read ... they are meant to be scored... Executive summaries are key and also the callout boxes, tables, graphs—things that an evaluator can copy and paste into their notes ... something like bolding or italicizing. Something to draw their attention to it. The overall format of the proposal is important, the way it looks, the graphics ... So, build that graphic library ... That's going to really move your proposal up there.

Your bid's aesthetics provide at-a-glance demonstrative evidence of your bid's quality. The overall design of your materials quickly communicates to evaluators how much you invested (and value) this opportunity and your solution. A professionally designed proposal implies that your customer and their needs matter to you. If you say you are the best of the best, then your delivery better look and feel like you are. For example, which looks and feels more professional—A or B (see Figure 14.2)?

Your page layout and graphics are not the only factors that make an impact. Your covers also quickly communicate volumes about your solution. In fact, three Air Force evaluators I met would jokingly bet on which solution provider was going to win based on the quality and efficacy of their proposal covers. Experienced reviewers have admitted to me that they could guess who had submitted the best proposal solely based on the covers. Their initial reaction—negative or positive—impacted the way they viewed and reviewed those bids.

Part of a cover's goal (see Figure 14.3) is to establish that you understand your customer's needs and you are a supportive, trustworthy, and reliable solution provider. If you can convey these elements on your cover, the government evaluator is more likely to be in a positive, agreeable state of mind when scoring your proposal. Emotions influence the very mechanisms of rational thinking [1, 2], so if the evaluator likes what they first see when receiving your proposal (i.e., the cover), then they are more likely to view your content through that positive lens.

Larry Tracy, who trained corporate executives to make oral presentations for government contracts, headed the Pentagon's top briefing team and worked for years with the Department of State. He was aware that design was so influential in the government's decision to purchase goods and services that it led to the government, at times, putting constraints on graphics by requiring black-and-white submissions or even requiring that no visuals be used to reduce the likelihood of influencing evaluators.

Throughout my years analyzing and working in the proposal industry, I found that the higher the bid award value, the more a company prioritizes graphic development. Our industry understands the influence that visuals have on their audience. It is common knowledge

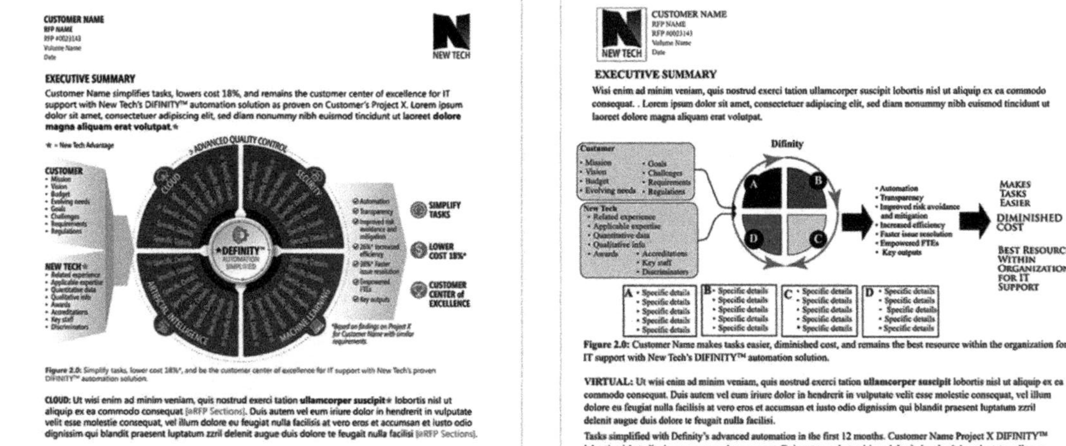

A

CUSTOMER NAME
RFP NAME
RFP #0023143
Volume Name
Date

NEW TECH

EXECUTIVE SUMMARY

Customer Name simplifies tasks, lowers cost 18%, and remains the customer center of excellence for IT support with New Tech's DIFINITY™ automation solution as proven on Customer's Project X. Lorem ipsum dolor sit amet, consectetuer adipiscing elit, sed diam nonummy nibh euismod tincidunt ut laoreet **dolore magna aliquam erat volutpat.**★

★ = New Tech Advantage

Figure 2.0: Simplify tasks, lower cost 18%*, and be the customer center of excellence for IT support with New Tech's proven DIFINITY™ automation solution.

CLOUD: Ut wisi enim ad minim veniam, quis nostrud exerci tation **ullamcorper suscipit**★ lobortis nisl ut aliquip ex ea commodo consequat [@RFP Sections]. Duis autem vel eum iriure dolor in hendrerit in vulputate velit esse molestie consequat, vel illum dolore eu feugiat nulla facilisis at vero eros et accumsan et iusto odio dignissim qui blandit praesent luptatum zzril delenit augue duis dolore te feugait nulla facilisi [@RFP Sections].

92% **OF TASKS SIMPLIFIED WITH DEFINITY'S ADVANCED AUTOMATION IN THE FIRST 12 MONTHS.** | Customer Name Project X DIFINITY™ lobortis nisl ut aliquip ex ea commodo consequat. Duis autem vel eum iriure dolor in hendrerit in vulputate velit esse molestie consequat, vel illum dolore eu feugiat nulla facilisis at vero eros et accumsan et iusto odio dignissim qui blandit praesent luptatum zzril delenit

Ut wisi enim ad minim veniam, quis nostrud exerci tation lobortis nisl ut aliquip ex ea commodo consequat Duis autem vel eum iriure dolor in hendrerit in exceeds requirements consequat, vel illum dolore eu feugiat nulla facilisis at vero eros et accumsan et iusto odio dignissim. Our current cloud performance metrics exceed requirements and **add a layer of advanced quality control**★ [@RFP Sections]...

CLOUD PERFORMANCE METRICS:
IOPS: 60,000
LATENCY: 1ms
MTBF: 2,000,000 hours

Quis nostrud exerci tation lobortis nisl ut aliquip ex ea commodo consequat. Duis autem vel eum iriure dolor in hendrerit in exceeds requirements consequat, vel illum dolore eu feugiat nulla facilisis at vero eros et accumsan et iusto odio dignissim. Lorem ipsum dolor sit amet, consectetuer adipiscing elit, sed diam nonummy nibh euismod tincidunt ut laoreet. Quis nostrud exerci tation lobortis nisl ut aliquip ex ea commodo consequat. Duis autem vel eum iriure dolor in hendrerit in consequat, **exceeds requirements** illum dolore eu feugiat nulla facilisis at vero eros et accumsan et iusto odio dignissim.

Page 2 of 18

©2021 Mike Parkinson and 24 Hour Company. All rights reserved.

B

NEW TECH

CUSTOMER NAME
RFP NAME
RFP #0023143
Volume Name
Date

EXECUTIVE SUMMARY

Wisi enim ad minim veniam, quis nostrud exerci tation ullamcorper suscipit lobortis nisl ut aliquip ex ea commodo consequat. . Lorem ipsum dolor sit amet, consectetuer adipiscing elit, sed diam nonummy nibh euismod tincidunt ut laoreet dolore magna aliquam erat volutpat.

Figure 2.0: Customer Name makes tasks easier, diminished cost, and remains the best resource within the organization for IT support with New Tech's DIFINITY™ automation solution.

VIRTUAL: Ut wisi enim ad minim veniam, quis nostrud exerci tation **ullamcorper suscipit** lobortis nisl ut aliquip ex ea commodo consequat. Duis autem vel eum iriure dolor in hendrerit in vulputate velit esse molestie consequat, vel illum dolore eu feugiat nulla facilisis at vero eros et accumsan et iusto odio dignissim qui blandit praesent luptatum zzril delenit augue duis dolore te feugait nulla facilisi.

Tasks simplified with Definity's advanced automation in the first 12 months. Customer Name Project X DIFINITY™ lobortis nisl ut aliquip ex ea commodo consequat. Duis autem vel eum iriure dolor in hendrerit in vulputate velit esse molestie consequat, vel illum dolore eu feugiat nulla facilisis at vero eros et accumsan et iusto odio dignissim qui blandit praesent luptatum zzril delenit

Ut wisi enim ad minim veniam, quis nostrud exerci tation lobortis nisl ut aliquip ex ea commodo consequat Duis autem vel eum iriure dolor in hendrerit in exceeds requirements consequat, vel illum dolore eu feugiat nulla facilisis at vero eros et accumsan et iusto odio dignissim...

SECURITY : Quis nostrud exerci tation lobortis nisl ut aliquip ex ea commodo consequat Duis autem vel eum iriure dolor in hendrerit in exceeds requirements consequat, vel illum dolore eu feugiat nulla facilisis at vero eros et accumsan et iusto odio dignissim. Lorem ipsum dolor sit amet, consectetuer adipiscing elit, sed diam nonummy nibh euismod tincidunt ut laoreet.

Page 2

©2021 Mike Parkinson and 24 Hour Company. All rights reserved.

Figure 14.2 Comparison of two renderings of the same proposal page.

Figure 14.3 Example of a high-impact cover for a Marine Corps proposal.

to companies like Northrop Grumman, Raytheon, Boeing, and Lockheed Martin that graphics are an essential part of winning new government business. It is not uncommon, when exceptional graphics are used, for government evaluators to commend the solution provider on their use of graphics.

I am not asserting that graphic communication is better than text. The combination of graphics and words has a communicative power that neither possesses singularly. "Pictures interact with text to produce levels of comprehension and memory that can exceed what is produced by text alone" [3].

Without graphics, an idea may be lost in a sea of words. Without words, a graphic may be lost to ambiguity. Robert E. Horn, an award-winning scholar at Stanford University's Center for the Study of Language and Information (CSLI), said, "When words and visual elements are closely entwined, we create something new and we augment our communal intelligence ... visual language has the potential for increasing 'human bandwidth'—the capacity to take in, comprehend, and more efficiently synthesize large amounts of new information."[5]

Like excellent writing, visual communication—when done the right way—improves understanding, recollection, and win rates. Good graphics explain your solution quickly and why your customer should choose you. Make the evaluator's job easier. Use professional visuals in your next proposal.

14.3 WEBSITES—PROPOSAL SUPPORT PLATFORM AND SOURCE FOR PROTEST

According to the Washington, D.C.-based independent research firm, Market Connections, Inc., "Websites are the #1 most important resource at various stages of the government buyer decision-making process" [4]. So, fielding and sustaining a company website is critical to your business success in the federal government marketspace.

Invest the time and resources to address the dimensions presented in Table 14.1. Pursue an internet marketing strategy of search engine optimization (SEO) to make your site better for search engines, such as Google, Bing, DuckDuckGo, and Yahoo! A key goal is to have your website rank high on the search engine results page (SERP).

Responsive web design (RWD) is a design approach to provide custom layouts to make webpages that render well on all screen sizes and resolutions while ensuring good usability. Certainly consider desktop and mobile versions of your website [6]. Consider that your organization's website is only 1 of 1.98 billion websites online [7]. Make it positively memorable.

Ms. Jennifer Schaus, principal at Schaus & Associates, recommended that U.S. government contractors build their website to clearly convey their core capabilities, federal contract vehicles, corporate

5. From: Boone, A., "Why Visuals Are So Important for Your Presentation," https://ethos3.com/why-visuals-are-so-important-for-your-presentation.

Table 14.1
Three Dimensions That Contribute to Successful Company Websites

Dimension	Guidelines	Tools and Outcomes
Appearance	Build high level of credibility through professional appearance of the website	Visualization tools: images, graphics, icons, animation, color
	Determine a well-balanced equilibrium between the aesthetic appearance and download speed	
Navigation	Ensure simple navigation	Sitemap
	Allow users to control where they are in every moment during the navigation and exactly what they can do	Backward button
Content	Organize and manage the site contents in a clear way to ensure ease of information searches	Information: up-to-date, relevant, comprehensible
	Provide high-quality, updated information and content	

Adapted from page 15 of [5].

certifications (e.g., ISO 9001:2015), and past performance, along with positive feedback from government customers, memberships in well-recognized industry associations, and technical white papers [8].

To that, I would add highlighting what it is that your organization is actually selling to government agencies. What outcomes are you offering [9]? How does what your firm bring value to the government [10]? What solutions do your company provide to help the government accomplish its mission, solve real-world problems, save time, or conserve discretionary dollars [11]? Finally, how will your firm serve your federal customer [12]?

14.4 PROPOSAL PROTEST SOURCE MATERIALS

The key to keep in mind is that your company's website may provide source material for a proposal protest. In one such Government Accountability Office (GAO) protest decision from May 2022 that involved the Department of the Army, the protester asserted that industrial training workshops conducted by the awardee's alleged subcontractor created an organizational conflict of interest. This information was found at the awardee's subcontractor's website. In another GAO protest decision from December 2022 that involved the U.S. Department of Health and Human Services (DHHS), Centers for

Medicare & Medicaid Services (CMS), arguments that the protestor made pertaining to Capability Maturity Model Integration certification, and SAFe principles relied on content from the winner's website.

As documented in a third GAO protest decision, also from December 2022 and involving the U.S. Army Materiel Command (AMC), the protestor argued that job postings on the winner's website demonstrated that the awardee knew it was unable to fill two key personnel positions required by the Task Order Request for Proposal (TORFP). In a fourth example, that being a January 2021 GAO protest decision involving the U.S. Marine Corps, the protester argued that the awardee's technical quotation could not have been properly rated as outstanding under the technical factor because a review of the awardee's website showed no recent and relevant experience with distribution management support services.

I have been involved personally in contesting a protest for a $250 million opportunity let through NASA Goddard Space Flight Center. The protester argued that the prime contractor who was awarded the contract did not show locations on its website within the required geographic proximity to Goddard. In addition, I have also seen situations in which a protest was brought against a small business because that business had inflated its employee count at its website for marketing and teaming reasons. The protestor argued that the small business had exceeded the size standard established by the U.S. Small Business Administration (SBA) for the North American Industry Classification System (NAICS) code applicable to the specific procurement.

Website accuracy is linked directly to successful proposal outcomes. Be sure that your company's website is completely accurate and comprehensive. It is being watched by your competition.

14.5 LINKEDIN PROFILES OF KEY PERSONNEL

Expect that government evaluators will search for your company's proposed key personnel on platforms such as LinkedIn. As the prime contractor, your company will be well-served to ensure that key personnel from inside your organization and from among your teammates refresh their LinkedIn profiles. Let's begin with the photo that appears on an individual's LinkedIn profile [13]. First, make sure that there is a photo rather than the faceless icon. The key person should select a photo that was taken within the past year and reflects how the person looks in a business or professional setting. I see far too many photos at LinkedIn that were taken at formal parties, in tradeshow environments, or on vacation. Frequently these images include people in addition to the individual whose profile is being highlighted, and often there are alcoholic drinks in hand. The ideal size for a LinkedIn profile photo is 400 × 400 pixels, which is 100 dots per inch (DPI) resolution, and it should be taken by someone other than the primary subject. In addition, the background should not distract from the key person.

All key personnel should update the "About," "Experience," "Education," and "Licenses & Certifications" sections of their profiles. They should ensure alignment between their LinkedIn profile and their proposal résumé. Further, to demonstrate thought leadership, focus attention on "Publications" [14]. This will bolster the individual's professional credibility.

References

[1] Bobrow, D., and D. Norman, "Some Principles of Memory Schemata," in D. Bobrow and A. Collins (eds.), *Representation and Understanding: Studies in Cognitive Science,* Academic Press, 1975, pp. 131–149.

[2] Rumelhart, D., "Schemata: The Building Blocks of Cognition," in R. J. Spiro, B. C. Bruce, and W. F. Brewer (eds.), *Theoretical Issues in Reading Comprehension: Perspectives from Cognitive Psychology, Linguistics, Artificial Intelligence, and Education,* New York: Lawrence Erlbaum Associates, Inc., 1980, pp. 33–58.

[3] Levin, J. R., "A Transfer of Appropriate Processing Perspective of Pictures in Prose," in H. Mandl and J. R. Levin (eds.), *Knowledge Acquisition from Text and Prose,* New York: Elsevier Science Publishers, 1989.

[4] Ocean 5 Strategies, "Government Contractor Websites—5 Essential Design and Development Considerations," https://www.ocean5strategies.com/government-contractor-websites-5-essential-design-and-development-considerations/.

[5] Flavián, C., C. Orús, and R. Gurrea, "Web Design: A Key Factor for the Website Success," *Journal of Systems and Information Technology,* May 2009.

[6] Cazañas, A., and E. Parra, "Strategies for Mobile Web Design," *Enfoque UTE,* V.7-Sup.1, 2017, pp. 344–357, http://ingenieria.ute.edu.ec/enfoqueute/.

[7] Minaev, A., "Internet Statistics 2023: Facts You Need-to-Know," December 26, 2022, https://firstsiteguide.com/internet-stats/.

[8] Schaus, J., "Marketing Basics for US Federal Government Contractors," April 19, 2022, https://www.youtube.com/watch?v=QWJI_Qiac9w.

[9] Frey, R. S., "Stop Selling Your Product! Build Tangible and Intangible Benefits into Your Proposals," *Texas Small Business News,* Vol. 4, Issues 6/7, June/July 2002, pp. 7–8.

[10] Gittens, A. M., "No One's Buying Your Products & Services. Stop Selling Them," LinkedIn, March 27, 2015, https://www.linkedin.com/pulse/ones-buying-your-products-services-stop-selling-them-gittens/.

[11] Kamps, H. J., "You're Not Selling a Product. You're Selling a Solution," August 23, 2017, https://medium.com/pitch-perfect/sell-the-solution-df62981ec51d.

[12] Robbins, Z., "5 Ways to Stop Selling (and Start Serving) Your Customers. It's Time to Turn Your Sales Pitch into a Satisfaction Strategy," *Inc.,* April 13, 2016, https://www.inc.com/young-entrepreneur-council/5-ways-to-stop-selling-and-start-serving-your-customers.html.

[13] Abbot, L., "10 Tips for Taking a Professional LinkedIn Profile Photo," LinkedIn, April 19, 2022, https://www.linkedin.com/business/talent/blog/product-tips/tips-for-taking-professional-linkedin-profile-pictures.

[14] Deehan, J., "20 Steps to a Better LinkedIn Profile in 2023," LinkedIn, January 21, 2023, https://www.linkedin.com/business/sales/blog/profile-best-practices/17-steps-to-a-better-linkedin-profile-in-2017.

15

PROPOSAL WRITING

15.1 PROPOSALS ARE KNOWLEDGE-BASED SALES DOCUMENTS

The essence of effective proposal writing is responding to the RFP or RFQ requirements while completing the process begun during marketing—namely, convincing the customer that your team and approach are the most beneficial and cost-effective. Remember that proposals are, first and foremost, sales documents through which validated strengths to the government are conveyed in terms they view as important. These span quality, schedule, cost, and risk mitigation, as well as security and safety. Proposals are not technical monographs, white[1] or position papers, or user's manuals. Conceptualizing and developing the technical solution is an information technology, engineering, or scientific issue. Packaging that solution in a combination of crisp, convincing narrative and high-impact graphics is a sales issue.

1. According to Stanford Law School (https://law.stanford.edu/wp-content/uploads/2015/04/Definitions-of-White-Papers-Briefing-Books-Memos-2.pdf), a "white paper is an authoritative report or guide that often addresses issues and how to solve them. The term originated when government papers were coded by color to indicate distribution, with white designated for public access."

Your company's proposal manager must articulate clearly to her writing team exactly what the expectations and acceptance criteria are. She must foster ongoing communication and feedback not only to herself but also among all of the writers. This will help to ensure a consistent approach, and importantly, minimize rework and rewriting. Driving down the level of rework aligns with a Lean business model. Less rework also translates to lower bid and proposal (B&P) costs.

Most proposals for a given procurement look and read essentially the same. The challenge is to incorporate well-substantiated information in your proposal that only your company can say. Identify precisely what will separate, or discriminate, your company from the competition. If your company is the incumbent contractor for a particular project and has performed favorably, use the names (and photos) of your incumbent staff people throughout your proposal. Reassure the customers that they will be working again with the same competent technical staff who will be supervised by the same responsive management team whom they know and trust. Help the government to understand that their investment in training and building intellectual capital will be retained. Write to a level of technical detail that exceeds what nonincumbents could gain by obtaining monthly technical progress reports through the FOIA and marketing outreach conversations with government decision-makers and influencers for the particular procurement. Demonstrate that your company understands the potential technical and programmatic risks and opportunities, as well as the success criteria. Strive to have your proposals not be boring.

15.2 ABCS OF PROPOSAL WRITING

Proposal content providers, writers, and editors should keep in mind the following ABCs of proposal writing:

> *A: Accuracy*—It is not enough, for example, to tell your customer that your company has saved time and money on projects of similar size, contract type, technical scope, and geographic footprint. Tell your customer exactly how much time you have saved, how you accomplished this, precisely how much mon-

ey was saved over what period of time, and the reasons why. Authenticate your claims with concrete, accurate information. Quantify your claims whenever possible with reference to your company's Contractor Performance Assessment Reporting System (CPARS) reports or Award Fee letters prepared by your federal government customers.

B: Brevity—Keep your writing crisp. Do not assume that the volume of words will make up for quality. Too many times I have heard technical staff say, "Give them 50 pages, and they'll just assume the correct responses are in there." Guess what? No, they (the evaluators) will not. Unless your company demonstrates through subheadings, graphical pointers, volume-level or sectional tables of contents, and cross-reference matrices that it has responded completely and appropriately to all elements in Sections L, M, and C, you will not score the maximum points available (or receive the highest adjectival or colorimetric rating).

C: Clarity—Assist the government evaluators to understand, for example, exactly how your company proposes to manage this new project, how you will integrate your subcontractor staff on the job, how your proposed project manager can access corporate resources quickly, and how you will manage tasks in geographically diverse locations. Amplify your crisp, direct narrative descriptions with clear, easily interpretable, and meaningful graphics. Graphics should have a visual entry point and exit point, with clear traceability in between.

Five pages of tightly integrated, well-constructed, and compliant narrative is much preferred to 20 pages that contain most every technical detail the writers happened to know. Technical data dumps glued together produce a very uneven and most likely noncompliant proposal. There is a natural tendency for content providers to dwell on familiar ground. Often, this is of the least significance as far as the RFP or RFQ requirements are concerned. Demonstrate how the features of your approach translate into performance, schedule, cost, risk mitigation, security, safety, and business-level benefits. Use facts—support all statements with concrete, quantified examples. Include the following:

- What your company will do for the customer;
- How your staff will do it;
- Why you will do it that way;
- What you did in the past, that is, your previous or current contractual experience. Highlight the relevant lessons learned, key practices discovered, and quantified successes generated from similar contractual experience.

Keep your ideas focused on the customer's requirements as stated in the RFP or RFQ, and keep your sentences short. I have seen sentences in proposal narrative that have been 85 words with no punctuation marks other than a period. Target your sentences to be in the 20 to 30-word range, as documented in the *American Educational Research Journal, Educational Theory,* and *Educational Horizons.* Above that quantity of words, comprehensibility and readability go down sharply. That being said, vary the sentence length to some extent to keep evaluators' interest. "Reserve the short sentences for main points and use longer sentences for supporting points that clarify or explain cause and effect relationships."[2]

Focus on translating the features of your technical and management approaches into benefits to the government and its project, its budget, its schedule, and its mission, business, and technology goals. This is an aspect of proposal writing that demands concrete, current, and in-depth marketing intelligence and sharing of this knowledge internally from business development staff to the capture and proposal teams. Without this, your proposal will describe your staff, facilities, contractual experience, and technical capabilities in a vacuum. There will be little or no direct linkage of your company's capabilities and solutions with your customer's perceived requirements and success criteria.

15.3 WRITING STANDARDS

There are several writing, editorial, and proofreading standards recognized by the federal government. Among these are:

2. Last, S., *Technical Writing Essentials: Introduction to Professional Communications in the Technical Fields,* University of Victoria, 2019, p. 45.

- U.S. Government Publishing Office (GPO), *Style Manual: An Official Guide to the Form and Style of Federal Government Publishing,* January 12, 2017.
- *The Tongue and Quill,* Air Force Handbook (AFH) 33-337, May 27, 2015. Resource for writing and speaking for the Army, Air Force, Marines, Navy, and Coast Guard.
- Sabin, W. A., *The Gregg Reference Manual: A Manual of Style, Grammar, Usage, and Formatting Tribute Edition,* 11th Edition.

Additional source of writing guidance:

- Last, S., *Technical Writing Essentials: Introduction to Professional Communications in the Technical Fields,* University of Victoria, 2019.

15.4 ABSTRACTING AS A PROPOSAL SKILL

In the 1980s, I worked as an indexer/abstractor at the NASA Scientific & Technical Information Facility (STIF) in central Maryland. Every workday, I, along with a group of other professionals, reviewed, read, and characterized at least 20 NASA, Department of Defense, Department of Energy, and/or academic books, published articles, reports, and papers across a wide spectrum of topics and disciplines. Topics ranged from aeronautics and plasma physics to advanced mathematics and government policies. Some documents were several pages, while others were hundreds of pages.

In characterizing the documents, we selected major and minor indexing terms from the *NASA Thesaurus,* a standard taxonomy or hierarchical vocabulary consisting of tens of thousands of terms and codified interrelationships. This was a large, bound volume that presented nested sets of terms that were related. The goal for the indexer/abstractor was to select the most applicable terms for each publication. Beyond that, we wrote 150-word abstracts of the documents on hardcopy NCR (no carbon required) forms, which were then keyed manually into the NASA/RECON online access system. RECON allowed users at NASA centers to order copies of documents electronically. NASA STIF also published an announcement journal, *Scientific and Technical Aerospace Reports* (STAR), twice each month. Of note is that abstracts and sets of indexing terms were quality-checked on an ongoing basis.

Relevant to proposal writing, the important skill that I learned through this experience was how to rapidly assess a document and prepare an accurate, cogent, and concise abstract. This would assist NASA engineers, scientists, and policymakers in choosing whether to invest their valuable time to request the full electronic file of a given source document. The abstract provides more information than that which is captured in the title of the document. It is a highly structured and coherent piece of writing that helps readers to understand the topics, focus, and conclusions of the larger document. Abstracting is a talent that directly supports crafting 5 to 8-sentence elevator speeches and main topic paragraphs in proposals. It enables one to capture the essence of what needs to be conveyed to the government evaluators or commercial customer.[3]

15.5 ACTIVE VOICE ADDS STRENGTH AND SAVES SPACE

Proposal writers should make every effort to employ the active voice: "We accomplished XYZ," not "XYZ was accomplished by us." The active voice adds strength to the proposal presentation and can save as much as 14% in terms of horizontal space as compared with passive voice, depending upon the font that is used. Importantly, active voice makes clear who is conducting the action. For example, "Interim deliverables were reviewed on a scheduled basis." Government evaluators would come away not knowing if the agency conducted the review, or if the contractor did so. Remove the ambiguity by using active voice: "Company XYZ's corporate quality assonance manager (QAM) will support our task leads with reviewing all interim deliverables prior to timely submission to the government for approval."

Contributors should attempt to vary sentence and paragraph structure. If you examine a proposal and most of the paragraphs on a page begin with "The" or your company's name, the narrative requires

3. Cremmins, E. T., *The Art of Abstracting,* 2nd ed., Info Resources Press, 1996. Cremmins' work is exceptional in helping one to understand how to convey information accurately and meaningfully while simultaneously conserving the number of words used.

editing to infuse variety. This same guidance applies to company marketing collateral and websites.

Writers should also try to employ strong and descriptive verbs, adjectives, and adverbs, such as the following:

- Our PM and team lead will use a RACI (responsible, accountable, consulted, and informed) matrix to help control work results and ensure proper engagement with each stakeholder for a particular work effort. *(32 words)*
- Our staff of six seasoned engineering professionals on the ground today in Huntsville all exceed the stipulated experience-level requirements by 2 years. *(22 words)*
- Our single-vendor contract management approach eliminates program accountability complexities, reduces costs, and minimizes overall management risks. *(16 words)*
- Our approach virtually eliminates risk to mission continuity during transition, achieves full staffing, and supports full contract operations by end of the 90-day phase-in. *(24 words)*
- Our risk management plan (RMP) will help create value (i.e., resources expended to mitigate risk will be less than the consequence of inaction), while ensuring that risk management is an integral part of our organizational rhythm. *(36 words)*
- We are performing outreach to incumbents right now to identify interest, skills, education, certifications, and qualifications. *(16 words)*
- To provide effective enterprise-level management of a geographically dispersed, multicustomer, and multiprogram work effort, our general approach for work management uses a Scaled Agile Framework (SAFe)-based management framework. *(28 words)*
- We performed a six-step resource analysis to assess the impact of cost on force modernization of critical Army weapons systems. *(20 words)*
- The sprint planning meeting, task breakout, and sprint cycle inform our program manager's and deputy program manager's assignment of resources. *(20 words)*

- This creates efficiency by using one system for job descriptions, applicant tracking, and each stage of the hiring process. *(19 words)*
- We proactively manage, track, report, and assess our performance in accordance with Department of Homeland Security-defined performance standards. *(18 words)*
- Our overall approach is based on the PMBOK 7th Edition that provides good practices for successful project life-cycle management. *(24 words)*
- We will separate tasks into meaningful work elements (sprint cycles), and apply the requisite people, processes, knowledge, and tools to accomplish the required work. *(24 words)*

Summarize the content of each proposal section in the first paragraph of discussion. Write for a variety of readers, including the skim or executive-level reader. To be a winner, a proposal must contain concise, understandable, and closely related thoughts woven into a compelling story. Identify the critical technical areas. In discussing them, use a level of detail that exceeds what a nonincumbent could use, and do not drown the evaluator in jargon. Your proposal should be built on solution-oriented, strengths-focused writing. Write-ups should be risk-aware and solution-oriented and demonstrate an understanding of the evolutionary changes likely to occur over the life of the contract. This is particularly true with today's 5-year, 7-year, and 10-year contract vehicles. Actually, in 2022, the U.S. Air Force awarded a $12 billion, 18-year contract called the Integration Support Contract (ISC) 2.0. This contract involves supporting the constantly evolving battle rhythm of quarterly objectives and key results (OKRs) in an essentially unknown and unprecedentedly complex operational world.

Make your company's compliance with the requirements and responsiveness to your customer's evolving needs apparent to the evaluator over and over again. Use RFP or RFQ terminology exactly (or in shortened form) for your proposal headings and subheadings. Employ the solicitation's terminology as a point of departure for further original writing. Do not simply recite the RFP verbiage in the actual narrative of your proposal. A Department of Defense Contracting Officer's Technical Representative (COTR) once told me that he felt personally insulted when contractors, both large and small, merely replaced the words "the contractor shall" that appear in the SOW, with "our company will." Such an approach demonstrates no technical understanding whatsoever. In a NASA RFQ, the government stipulated that "[s]tating that you understand and will comply with the specifications, or paraphrasing the specifications is inadequate, as are phrases such as 'standard procedures will be employed' and 'well known type techniques will be used.'"

Define acronyms and abbreviations the first time they are used in each major proposal section. Not all evaluators see the entire technical, management, and past performance proposal volumes, so redefinition of acronyms is certainly acceptable assuming page-count constraints allow. Avoid sectional or page references in the text because references must be changed each time the section or page numbers change. References such as "See Section 5.1.1 on page 5–34" should not be used. Let the compliance matrix and table of contents assist the evaluators in locating specific pages or sections. In addition, avoid fifth-order headings (for example, Section 4.3.1.3.1).

Proposal writers should attempt to think graphically as well as in written form. Every proposal section should have a figure or table associated with it that is referenced and discussed clearly in the text. Figure captions should appear centered beneath the appropriate figure. Table legends should appear centered above the appropriate table. Assuming the RFP or RFQ permits, label all illustrations as "exhibits" to avoid confusion between what is a table and what is a figure. I have witnessed many debates about determining whether to call something a table or a figure. No need to expend energy in this regard.

15.6 ACTION CAPTIONS

The term "action caption" originated with Mr. Hyman Silver, an icon in the aerospace industry and developer of The Hy Silver Proposal Writing Method. A great place to reinforce sales messages is in the captions or legends for figures, tables, photographs, and other graphical presentations. Let's look at the difference between simply identifying what a graphic or photograph is, and using the caption as a forum to convey important sales messages that have been tailored to the specific procurement.

- Instead of "Project Organizational Chart," your caption might read: "Our project team leads are proficient in each of the USACE technical requirement areas."
- Instead of "Map of Our Local Locations," your caption might read: "The proximity of our offices to Fort Sam Houston Army Base will enable rapid response to changing requirements."
- Instead of "Capabilities Matrix," your caption might read: "Each one of our four senior technical staff holds an accredited degree in a relevant engineering and scientific discipline."
- Instead of "Our Laboratory Staff," your caption might read: "Our well-trained laboratory staff are dedicated to and trained in quality-oriented procedures that ensure appropriate chain-of-custody handling."
- Instead of "Appropriate Health and Safety Procedures," your caption might read: "We actively practice a comprehensive health and safety program designed to meet the requirements of 29 CFR 1926.65 for hazardous waste site work."
- Instead of "Scope of Support Services," your caption might read: "We provide cost-effective, innovative support services in support of the Dodd-Frank Wall Street Reform and Consumer Protection Act."

In each one of these examples, we have taken a flat statement and added positive, dynamic sales messages. We have introduced benefits of your approach and capabilities to the customer. These sales messages should be developed early in the proposal response cycle so that the entire proposal writing team can incorporate them into their narrative and into the captions that they develop for their graphics. Please note, however, that it is always critical to be totally compliant with the RFP. If a particular RFP stipulates that figure and tables shall be named in a particular manner, you must adhere to that guidance.

15.7 METHODS OF ENHANCING YOUR PROPOSAL WRITING AND EDITING

Toward the goal of keeping your proposal writing highly focused, consider the following four stages:

1. Collect the technical and marketing intelligence information that you need for your section(s). Focus on the basic features of those materials.
2. Identify the information in those materials that is relevant to the RFP requirements for your particular section and which supports the evidence of strengths that have been earmarked for that section.
3. Extract, organize, and reduce the relevant information. Begin to compose the relevant information into sentence form in accordance with the proposal outline that was part of the kickoff package or proposal directive. Slant your writing toward your audience, that is, the customer evaluators of the proposal. Be informative but brief. Be concise, exact, and unambiguous. Use short, complete sentences.
4. Refine, review, and edit the relevant information to ensure completeness, technical accuracy, and the inclusion of sales messages.

Importantly, one of the best ways to polish the writing in your company's proposals is to read it aloud to a colleague. Awkward sentence constructions become obvious quickly. The flow of the narrative can be smoothed and refined through this process of open reading.

A longer-term key to effective proposal writing is reading many different types of materials across a variety of disciplines and media—such as professional and scholarly journal articles, public relations brochures, online newspapers and business magazines, textbooks, novels, and other works of literary fiction. An enhanced functional vocabulary is one result. Another is learning how to use the same word in a broad spectrum of contexts. Reading in this manner can also produce an appreciation of nuance, that is, using precisely the most appropriate word for the given application.

In terms of editing proposal narrative, the following checklist can prove very useful:

1. Check for completeness.
2. Check for accuracy.
3. Check for unity and coherence.
4. Include effective transition statements.

5. Check for consistent point of view.
6. Emphasize main ideas.
7. Subordinate less important ideas.
8. Check for clarity; eliminate ambiguity.
9. Check for appropriate word choice (diction).
10. Eliminate jargon and buzzwords.
11. Replace abstract words with concrete words.
12. Be concise.
13. Use the active voice as much as possible.
14. Check for parallel structure (for example, a list begins with all *-ing* or all *-ed* words).
15. Check sentence construction and ensure sentence variety.
16. Eliminate awkward expressions.
17. Eliminate grammar problems.
18. Check for subject-verb agreement.
19. Check for proper case (such as upper case and initial caps).
20. Check for clear reference of indefinite pronouns.
21. Check for correct punctuation.
22. Check for correct spelling, abbreviations, acronym definitions, contractions (avoid them), italics, numbers, symbols, for example.
23. Check for correctness of format.

15.8 RESOURCES TO SUPPORT YOUR COMPANY'S PROPOSAL WRITING EFFORTS

- Brown, G. W., *Federal Government Proposal Writing: A Simplified Teaching for Beginners and Small Businesses,* CreateSpace Independent Publishing Platform, 2016.
- Sant, T., *Persuasive Business Proposals: Writing to Win More Customers, Clients, and Contracts, 3rd ed.,* New York: AMACOM (now part of HarperCollins Leadership), 2012.

15.9 POTPOURRI

When working in proposal land, take the time to adjust Microsoft Word's autocorrect feature so that "manager" does not become "manger."

The following is one of a set of real questions that emerged during several of my recent F2F proposal training seminars in Maryland and Ohio. My answer (A) follows the question (Q).

Q: How might we facilitate the writing of the nontechnical aspects of a proposal?

A: Many of these sections can be written and illustrated in advance, and then tailored. For example, your company's risk management plan, subcontractor management plan, organizational conflict of interest (OCI) plan, training and cross-training approaches, and retention successes can be crafted before you have an RFP staring you in the face. That is because they remain largely the same across bids.

15.10 TAKE OFF IN THE RIGHT DIRECTION: DEFINING ACRONYMS EARLY IN YOUR PROPOSAL DEVELOPMENT PROCESS

Most commonly, proposals developed in response to U.S. federal government RFPs are literally jam-packed with acronyms. A challenge for proposal professionals is ensuring that the meaning of a given acronym is accurate for the context in which it is used. For example, the acronym FTE has multiple meanings depending upon the situation—in a résumé (flight test engineer), a past performance citation (factory test equipment), a cost/price volume (full-time equivalent), or a technical approach (fault tolerant Ethernet).

Too many times, I have observed that organizations task an editor or desktop publishing (DTP) specialist with defining the myriad of acronyms that appear throughout the proposal volumes. Despite the best efforts of these individuals, this practice can result in acronyms being misdefined. This casts a negative light on the proposing organization (the offeror), and frankly, government evaluators deserve much better. There is no valid reason to introduce confusion into your proposal—it does not improve your probability of winning (P_{win}) whatsoever.

To be sure, there are multiple online sources that can assist non-SMEs with identifying at least candidate definitions of acronyms. These encompass the following:

- Defense Acquisition University Press, Glossary of Defense Acquisition Acronyms & Terms;
- DoD Acronyms Dictionary;
- DoD Cyber Exchange Public: Cybersecurity Acronyms: https://public.cyber.mil/acronyms/;
- Google Play Acronym Finder;
- MilitaryWords.com;
- NASA Acronyms and Abbreviations: https://www.nasa.gov/pdf/632702main_NASA_FY13_Budget-Reference-508.pdf;
- National Archives Library Information Center (ALIC): https://www.archives.gov/research/alic/reference/military/military-reference-materials.html;
- U.S. Department of Health and Human Services (DHHS) Office of the Assistant Secretary for Planning and Evaluation (ASPE) Common Acronyms: https://aspe.hhs.gov/common-acronyms;
- U.S. DoD Military Acronyms and Abbreviations: https://www.ehso.com/DOD-Acronyms-Military-Abbreviations.php.

However, I recommend an alternative. Beginning right at the kickoff meeting—whether virtual or F2F—the capture manager and proposal manager of the prime contractor should communicate the no-kidding expectation that defining acronyms in all proposal work products submitted is a prerequisite for them to be accepted by the proposal team. That message needs to be directed to both the prime and all subcontractors. Specifically, all acronyms need to be defined the first time that they appear in any proposal work product, whether it be a corporate overview, subcontracting plan, or key personnel résumé.

A second recommendation is for the proposal manager or proposal coordinator to build and maintain an updated master acronym list. This should be posted prominently at the knowledge-sharing platform that the team is using, whether that is Microsoft SharePoint, Confluence, Google Workplace, Microsoft Teams, or another tool.

Reminders should be issued as necessary at the regular tag-up meetings to reinforce the value of getting every acronym defined accurately for the context in which it appears in the proposal volumes.

Importantly, taking this approach saves a lot of time during the final production phase of the proposal life cycle (see Figure 15.1). Your organization will submit a better product for evaluation that does not contain linguistic mysteries.

15.11 DOVETAILING—FIT YOUR PROPOSAL EXACTINGLY WITH THE FEDERAL RFP

Recently, while supporting a civilian agency proposal valued at more than $100 million, a consulting firm who was engaged to assist the

Figure 15.1 Taking off in the right direction, Southwest Airlines' jet departs from San Diego International Airport. (Photograph © Dr. R. S. Frey.)

small business prime contractor, developed an outline for each of the four volumes, as they had been tasked to do. In the RFP, this particular government agency stipulated that: "The format for each proposal volume shall parallel, to the greatest extent possible, the format ... contained in Section L of this solicitation." In effect, an industry's outline must necessarily mirror the structure of Section L, Instructions, Conditions, and Notices to Offerors.

However, as happens all too frequently, the outline that was developed did not adhere to the structure of Section L. For example, in Section L, B.1.b, the following direction is provided: "The Offeror shall describe its teaming approach to include subcontract management." That should have led to the generation of an outline that included the subheading below: II.B.1.b: Teaming Approach/Subcontract Management.

To break this down into its components, the Roman numeral "II" denotes Volume II of this proposal. The "B.1.b" comes directly from Section L.

Somewhat surprisingly, the outline subheading for this specific topic was rendered as: MA1.2 TEAMING APPROACH/SUBCONTRACT MANAGEMENT.

First, this RFP did not reference teaming approach and subcontract management in ALL CAPS. Immediately, a new element has been introduced where it should not have been. Second, the provision in the RFP indicated that Section L must drive the format of industry's proposals. In addition, I do not understand the use of the Arabic numeral 2 in this subheading. It does not appear in the RFP in this context at all. There is some latitude with the reference to "MA" for management approach. But "MA1," if used at all, should have been "MA-1" with a hyphen between the "MA" and the "1," just like in the RFP.

The outline, as developed, will make the government evaluators work harder than they should in order to map the prime's proposal response to the government agency's RFP. That is not a good thing at all. Government evaluators already have a day job. They have multiple proposals to review and evaluate for a given acquisition. Do you think that they will be positively inclined toward a proposal that has introduced an unwelcome degree of complexity when none was asked for or required? The P_{win} just went down, and for no good reason at all.

Just like in woodworking and the incredibly strong dovetail joint used in drawers, make certain that your proposal fits tightly together with the provisions in the RFP (see Figure 15.2). It makes for a far stronger bid.

15.12 WORDS TO AVOID IN YOUR PROPOSALS

Think twice before using the following words and phrases in your next proposal to the U.S. federal government (see Figure 15.3). Help the government evaluators see your proposal as standing above the competition.

- "propose," "proposed": It is a proposal, no need to use the word "propose."
- "etc.": Takes up valuable space, and conveys nothing of value.

Figure 15.2 My handcrafted cherry and walnut wood end table with exacting dovetail joinery for the drawer. (Photograph © Dr. R. S. Frey.)

Figure 15.3 Vintage manual typewriters on display in The Press Hotel in downtown Portland, Maine. (Photograph © Dr. R. S. Frey.)

- "We believe": Rather, "our 4 years of experience supporting the FAA provides us with the insights to confidently assert that…."
- "many": Quantify
- "vast": Quantify
- "world-class": No small or medium-size business offers world-class anything.
- "unique": Very few processes, approaches, tools, and techniques are genuinely one-of-a-kind.
- "extensive": Quantify.
- "ahead of schedule": Quantify.
- "the PM": It is your organization's project manager; use "our PM," or better yet, name that individual. Take ownership.
- "uninterrupted," "disruption": Even if the RFP uses this language, change to "continuity of operations."

- "We understand": Instead, convey exactly what it is that your organization does understand.
- "Company ABC has a long history": Quantify.
- "ISO 9001:2015-certified": ISO does not certify anyone or anything. Instead, Company ABC's Quality Management System (QMS) is certified to the ISO 9001:2015 standard.
- "includes, but is not limited to": Avoid this kind of legalistic language.
- "is responsible for": Instead, use a strong action verb to convey what the person or organization actually does.

Now go win more stuff!

16

ACHIEVING PROPOSAL STRENGTHS THROUGH FORWARD-LOOKING BUSINESS DECISIONS

16.1 SOURCE SELECTION DOCUMENTS AND PROPOSAL STRENGTHS

NASA calls them source selection statements. The Department of Defense, Department of Energy, Department of Commerce, General Services Administration, Department of Homeland Security, and the United States Department of Agriculture refer to them as source selection decision documents. Either way, these statements and documents can offer meaningful information about what the particular government agency has determined to be strengths in competitive solicitations (see Figure 16.1).

So how do companies go about locating the source selection documents? First, every time that your firm bids on a competitive U.S. federal government competitive proposal, you will receive some type of source selection document. Second, these documents can be obtained through FOIA requests. Third, search the SAM.gov website

Figure 16.1 Preseason football practice with the Washington Commanders in Ashburn, Virginia. (Photograph © Dr. R. S. Frey.)

for "source selection statement." You can also conduct searches online through Google, DuckDuckGo, Bing, or Yahoo! for "source selection statement" and "PDF." Service providers such as Deltek also have specific Source Selection documents available as part of the paid subscription.

As we know in the federal competitive proposal arena, a critical goal for all offerors is to maximize the number of strengths and significant strengths that the government evaluators are able to locate in the proposals submitted. I have seen a single award winner have 38 strengths and the runner-up have 32 strengths. Remember that the U.S. federal government does not care about industry's win themes or discriminators. The focus is on strengths, weaknesses, deficiencies, cost/price, and past performance.

For example, in 2021, DHS had issued an RFQ to acquire "professional technical support services to assist the organizations, missions, functions, and objectives enabling DHS [Science and Technology

Directorate] S&T to integrate innovative technology into everyday use by the DHS operational components." In the DHS SSDD, one of the offerors earned a strength for their proposed program manager having 17.5 years of relevant experience with program, project, and portfolio management. The baseline level of experience had been 15 years. Another offeror earned a strength for its approach to software license management.

In an SSDD that the Army prepared for aircraft training services in Alabama, a strength was assigned to one offeror proposing that training and services would be housed in one location. That same firm also received strengths for innovation and creating an excellent twenty-first-century approach to flight training. Another offeror, as the incumbent, earned a strength for having "very few changes to make post award."

In a 2021 NASA Source Selection Statement for an operations-related contract, one offeror received a strength for its "project lead talent pipeline and a project management certification program...." Another offeror earned a strength for its "knowledge-sharing initiative."

Finally, the Department of Justice (DOJ) also uses SSDDs to capture its findings following proposal evaluation. One SSDD for an IT support services contract noted a strength for one of the offerors as having "trained, cleared, staff that is familiar with the existing systems and has relevant experience, educations, and skill sets."

Strengths are specific to a given proposal. It is counterproductive for companies to develop a battery of strengths that are then force-fit into proposals. For example, a large corporation based in the Mid-Atlantic region proposed using Capability Maturity Model Integration Maturity Level (ML) 5 processes to improve performance. Sounds logical, even forward-leaning. However, the government agency determined that the cost, documentation, and administrative requirements associated with ML 5 were overly burdensome. What is a strength in one proposal for one agency may not be a strength in another proposal for another agency. It may be judged to be a weakness.

Focus on maximizing the number of context-specific, quantitatively and qualitatively validated strengths that are aligned with the Evaluation Factors for Award presented in Section M. That should be Capture and Proposal Teams' #1 Goal.

16.2 LINKING KEY BUSINESS DECISIONS TO PROPOSAL STRENGTHS

What should key overhead staffing selections, capital investments, independent research and development (IRAD), and corporate and staff-level certifications and accreditations have in common? Each one of these important decisions ought to be made with an eye toward how it will potentially rise to the level of a strength in your organization's proposals to the U.S. federal government. Senior leadership would be wise to consider how these choices might become "[a]n aspect of the proposal that greatly enhances the potential for successful contract performance and/or that appreciably exceeds specified performance or capability requirements in a way that will be advantageous to the Government during contract performance" (NASA's definition of a strength; the DoD's definition is similar).

Let's look at key overhead staffing selections. The decision to invest in a highly qualified corporate technology officer (CTO) could contribute to effective customer liaisons (i.e., a strength) and forward-looking strategic planning. CTOs can leverage business intelligence tools to identify insights and then translate those insights into new services that meet government agency needs (i.e., a strength). As technology becomes more of a source of competitive advantage, the CTO's contribution becomes increasingly valuable. Or consider hiring a field-proven nationwide talent recruiter, one who is experienced in locating and attracting low-density/high-demand SMEs in such markets as Silicon Valley in California's Bay Area and the Silicon Slopes around Lehi, Utah. Given the importance of timely staffing on federal contracts, the right individual in this role with a track record of success certainly could contribute to an organization receiving a proposal strength under management approach.

Insofar as the allocation of capital investments and IRAD resources, organizations should consider the proposal value of Innovation Hubs, Technology Accelerators, eLabs, Centers of Excellence, Value Creation Centers, or MindLabs. When accompanied by photographs and success metrics that bring these assets to life, such investments can certainly rise to the level of a strength.

Investing the time, energy, and resources to have your organization's quality management system (QMS) certified in accordance with the International Organization for Standardization (ISO) 9001:2015 standard can certainly strengthen your proposals. It may also make business sense to manage information security in accordance with the ISO/ International Electrotechnical Commission (IEC) 27001 international standard. The Army, the GSA, and the Department of Veterans Affairs (VA) have put a premium on ISO quality certifications, as evidenced in numerous task orders and in the VA Veterans Technology Services (VETS) 2 Indefinite Delivery/Indefinite Quantity (ID/IQ) contract, the Army Information Technology Enterprise Solutions–3 Services (ITES-3S) IDIQ contract, and the GSA Alliant 2 Governmentwide Acquisition Contract (GWAC). In 2022, in a Civilian Agency Blanket Purchase Agreement (BPA), the government required that staff proposed under four different labor categories hold the Project Management Professional (PMP®) certification—as a minimum requirement. Choosing to invest in the PMP certification for program-level and project-level staff can spell the difference between being able to compete at all and winning the opportunity.

As part of your organization's strategic planning process, I strongly recommend that the abovementioned decisions be linked with how they can contribute to receiving strengths in upcoming federal proposals.

16.3 POINTING THE WAY—ANSWERS TO TRAINING PARTICIPANTS' QUESTIONS

The following is a real question that emerged during one of my recent F2F proposal training seminars in Maryland and Ohio. My answer (A) follows the question (Q).

Q: Why are discriminators not as important as strengths?

A: "Discriminators" are part of industry's lexicon. The federal government does not think in those terms, nor do they evaluate proposals using that terminology. This is evidenced from source selection statements, and the DoD's Acquisition guidance.

16.4 POST-AWARD DEBRIEFINGS

U.S. Army Post-Award Debriefings are conducted to instill confidence that all proposals were treated objectively; to highlight the strengths, deficiencies, weaknesses, and past performance information identified during evaluation; and to assist offerors to improve their proposals for future acquisitions. Typically, the Army assesses three factors during evaluation: (1) past performance, (2) mission capability, and (3) price.

Offerors' past performance submissions are assigned an adjectival evaluation rating—excellent, good, adequate, marginal, poor, and unknown. Excellent equates to a risk level of "very low."

The mission capability factor portion of offerors' proposals also receives an adjectival evaluation rating, ranging from excellent, good, and acceptable to marginal and unacceptable. As with past performance, an excellent mission capability evaluation rating translates to a risk level of very low. During its proposal evaluation process of mission capability, the Army identifies significant strengths, strengths, significant weaknesses, weaknesses, deficiencies, and uncertainties.

An example of a significant strength for one particular acquisition conducted through White Sands Missile Range (WSMR) was: "Proposal shows strong presence at other agencies and suggests substantial reachback capability. This further reduces risk since the offeror would take advantage of established processes and procedures and be effective at contract award." One of the strengths identified was: "[c]lear use of performance metrics will lower performance risks by ensuring measurable and reportable process performance." Among the weaknesses, the Army noted that for the "Staffing plan: percentage of entry level positions seems high, increases risk to contract performance and to offeror's goal of incumbent capture."

Finally, uncertainties included the finding that the: "[p]roposal relies heavily on performance and experience of subcontractor...."

Post-award debriefings, source selection statements, or SSDDs should definitely be mined for insights when pursuing business with U.S. government agencies.

16.5 LEVERAGING SOURCE SELECTION STATEMENTS TO MAKE KEY BUSINESS DECISIONS

One way to help your company make intelligent business and investment decisions related to a specific federal acquisition opportunity is to conduct an analysis of recent source selection statements or SSDDs from the same government agency. Document the significant strengths and strengths that the government agency identified through source selection, and leverage them to develop a listing of potential actions your firm could take on an upcoming procurement.

Let's take a look at several real-deal significant strengths and strengths from recent source selection documents issued by a specific civilian agency.

- *Significant strength:* Corporate investments, which included phase-in cost ... plan to provide all software to [the agency] at the end of the contract, and investments in [independent research and development] IR&D.
- *Strength* was for ... [a] knowledge-sharing initiative.
- *Strength* was for a project lead talent pipeline and a project management certification program.
- *Strength:* Financial commitment to employee career development through educational reimbursement, training courses, and conference attendance demonstrates an effective retention and training approach.
- *Strength:* Effective approach and implementation of the proposed functional area and improvement manager (FAIM) position.

Note the importance given to investments and financial commitments on the part of prime contractors. Meaningful and validated investments in support of the particular program frequently rise to the level of a strength. Identifying, training, certifying, and incentivizing human talent is another area frequently of especial importance to government evaluators. The government wants to ensure that the staff in whom they have invested time and training stays engaged on the new contract.

One action item to complete before the release of the final RFP is to conduct a thorough internal inventory of all of the dollars that your company has invested in taking care of your staff professionals over the past 3 years, for example. How much did your firm spend on employee educational reimbursement, technical refresh, certifications, and cross-training? How much was spent on employee engagement and morale-building events, as well as tangible awards, bonuses, and peer recognition?

In some cases, it may make sound business sense for your company to invest in a full-time or part-time individual to focus on such areas as improvements, innovations, knowledge sharing, or marketing outreach on the upcoming contract. Quantify the investment level to which your company will commit, and indicate the duration of that investment.

The bottom line is to use what you know a particular federal agency has cited as strengths in the recent past to inform your own company's business decisions related to a specific procurement with that same agency.

17

LOOKING AT PROPOSAL REVIEWS FROM DIFFERENT ANGLES

17.1 VERTICAL AND HORIZONTAL PROPOSAL REVIEWS

We should absolutely not get wrapped around the axle with how proposal reviews are named or designated (see Figure 17.1). For example, I have seen and participated in the Blue, Pink, Red, Gold, Silver, Yellow, and Green Teams. The Red Team in particular comes heavily freighted across the federal support services marketspace. Participants bring with them highly divergent viewpoints as to what areas of focus should be addressed during the Red Team. I have also worked on a proposal that involved a government-industry partnership, in which case the government participants had no grounding whatsoever in industry proposal review terminology. In that instance, we simply designated reviews as Review 1 and Review 2. The important thing is to define the focus and outcomes of each review to the members of the review team in advance of the actual review.

That brings me to my point. Organizations should employ what I call "vertical" and "horizontal" reviews (see Figure 17.2) and do both of them at each major review cycle. Focus the vertical review on:

Figure 17.1 Tools to measure and cut wood at various angles. (Photograph © Dr. R. S. Frey.)

(1) validating exacting compliance with the solicitation documents (including Section L, Statement of Work (SOW)/Performance Work Statement (PWS), Data Requirement Deliverables (DRDs), and Section M); (2) confirming the accuracy of the proposal structure (including adherence to the RFP framework, numbering or alphanumeric scheme in the solicitation (when the RFP uses B.1.a. through B.1.f., do not introduce a new outlining scheme of B.1.1 through B.1.6), and specific terminology (do not name a heading "End User Support" when the RFP uses "Deskside Support"); and (3) verify the accuracy and completeness of technical and programmatic narrative and graphics.

In the horizontal review, appoint a specific subset of reviewers to conduct a strengths assessment across all proposal volumes, confirming that clear, quantitatively substantiated evidence of strengths, which align with Section M Evaluation Factors for Award, are apparent

Figure 17.2 Overhead horizontal walkway in the foreground against the tall vertical Marriott Marquis Hotel in San Diego, California. (Photograph © Dr. R. S. Frey.)

throughout all proposal sections and volumes. In addition, verify consistency across such elements as key concepts, quantitative data, and position titles. Further, ensure that the entire proposal response presents a forward-looking view to meet the government agency's evolving mission set and operational environment. Contracts are dynamic. Our proposals need to reflect that fact.

17.2 MEASURE WHAT IS IMPORTANT—ENSURING THAT STRENGTHS FOR YOUR COMPANY APPEAR IN THE GOVERNMENT'S SOURCE SELECTION STATEMENT

To ensure that federal government evaluators ultimately codify your organization's "candidate" proposal strengths in their source selection statement for a given procurement, several key steps need to occur (see Figure 17.3). First, government evaluators must read relevant strengths in your company's written proposal or hear them articulated during delivery of your oral presentation. Importantly, to be considered strengths, the evidence that your company presents in this regard must resonate with what the government agency deems as adding value to their mission and to their program or project.

On a recent DoD proposal in which I was engaged as the oral presentation coach, the government did not allow PowerPoint slides or any other written materials to be presented. This meant that strengths had to be communicated verbally during the 60-minute F2F oral delivery; there was nothing for the evaluators to read during that time. Within DoD [1], a "strength is an aspect of an offeror's proposal with merit or will exceed specified performance or capability requirements to the advantage of the Government during contract performance."

For evaluators to read relevant strengths in your company's written submittal, they must find them easily within the narrative and graphics of the proposal documents. Strengths must "jump off the page." How is this done? One way is through the use of cogent, Section M-focused language. What your company puts forward as evidence of a candidate or potential strength must clearly demonstrate an "Outstanding" or "Exceptional" approach and understanding of the requirements (per the *DoD Source Selection Procedures* [1]). So, using the words "Outstanding," "Exceeds," "Exceptional," and

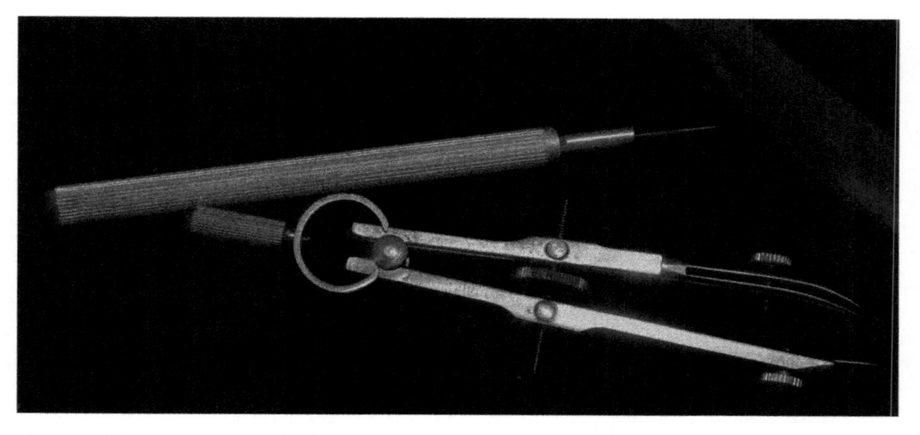

Figure 17.3 Precision drafting and measuring tools. (Photograph © Dr. R. S. Frey.)

"Advantageous" in the text, in callout boxes brought to life using colors that align with the government agency's color palette, and in figure/exhibit captions or table legends, is a very good thing to do. However, these keywords must be accompanied by tangible, quantitatively validated evidence to support these assertions. I always recommend to my customers to use the phrase "evidence of strengths" in their proposals, rather than merely "strengths." Frankly, the use of strengths alone amounts to an arrogant assertion. Strengths are in the eye of the government beholder (also known as the evaluator), not industry.

One technique that I had developed and implemented is for companies to track strengths as a key metric during their proposal review process (see Figure 17.4). I had the opportunity to present this technique formally to a *Fortune* 50 international defense contractor several years ago during a gathering of a select group of their senior staff from North America and Europe. My guidance was to have a subset of reviewers at each major review stage be charged with finding the strengths within the nonprice proposal volumes that the company identified during the capture management phase. The reviewers receive a listing of the candidate strengths; their task to actually find them in the proposal documents. Let's say that an organization has identified 25 candidate strengths for a given opportunity. Ideally, reviewers should be able to locate 100% of them. The metric to track at each review is the actual percentage of those candidate strengths that the reviewers find in the proposal. As the proposal matures, the percentage should definitely increase. The ultimate litmus test is

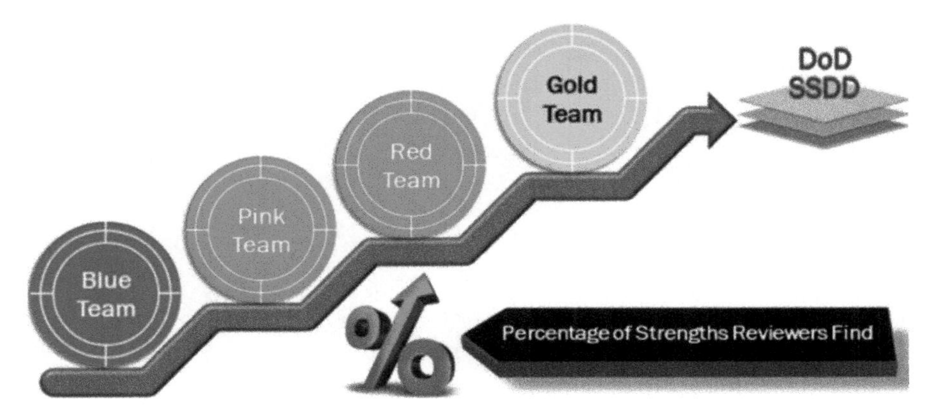

Figure 17.4 Key metric to track during proposal reviews.

determining the percentage of candidate strengths that your company developed and embedded in your proposal, which appear in the government's source selection document. On a NASA proposal that was hundreds of millions of dollars that I supported from 2019 to 2021, it was extremely gratifying to see candidate strengths that the prime contractor had built into the proposal appear almost verbatim in the source selection statement, and the proposal was a winner.

17.3 COMBINING AGILE SCRUM AND SAM

Whenever I grapple with putting concepts into a broader context, I think in terms of using a wide-angle (17–40-mm) camera lens versus a macro (100-mm) lens to take a photograph (see Figure 17.5). The wide-angle lens gives you a wide field of view (FOV). The wider the FOV, the more of the "scene" you will be able to see in the frame of the photo. The "scene" in question might actually be conceptual models. As noted by German computer scientist and professor of information systems engineering Bernhard Thalheim [2], *"Conceptual modelling*

Figure 17.5 Wide-angle camera lens. (Photograph © Dr. R. S. Frey.)

aims to create an abstract representation of the situation under investigation, or more precisely, the way users think about it." Among the purposes of a model are *"perception support* for understanding the application domain" [2].

Both the Successive Approximation Model (SAM) for instructional design (ID) and Scrum values, principles, and practices share the application domain of the Agile Manifesto. Developed by Dr. Michael W. Allen, SAM, with "its agile and iterative approaches" [3], is a cyclical process "focused on learner experiences, engagement, and motivation" [4]. The Agile Manifesto in the Scrum expression helps businesses to organize their work, such as proposal development, to maximize collaboration, deliver frequently, and create multiple opportunities to inspect and adapt (https://www.scrumalliance.org/about-scrum#!section3). With its roots in the Agile Manifesto, Agile Project Management has expanded well beyond software development and is now becoming a cutting-edge management approach applied in high-velocity competitive markets, and with rapidly changing technologies, innovation-driven customers, and high levels of uncertainty [5].

The PMBOK defines a "project" as a "temporary endeavor undertaken to create a unique product, service, or result." A proposal is, essentially, the "product" or "result" of a temporary endeavor. So with the convergence of the concepts of project and proposal, APM can be introduced to accelerate and increase the predictability of product (i.e., proposal) delivery, increase productivity, and enhance product quality, as well as improve project visibility, reduce project risk, and reduce project cost [5] by driving down the level of rework.

As shown in Figure 17.6, there are multiple connect points between SAM and Scrum as applied to proposal development. First, let's zero in on the SAM, which I have modified from Michael Allen's original workflow [6]. Borrowed from the field of Agile e-Learning development, and in a manner similar to Agile software development, the SAM in a proposal development context includes a preparation phase, an iterative proposal design phase, and iterative proposal development phase. The SAM, as adapted to the proposal world, is a fundamentally different approach to successful proposal development. What is the difference? The SAM addresses the proposal lifecycle activities through iterations and repeated small steps, rather than with giant, schedule-driven strides alone. Rapid development,

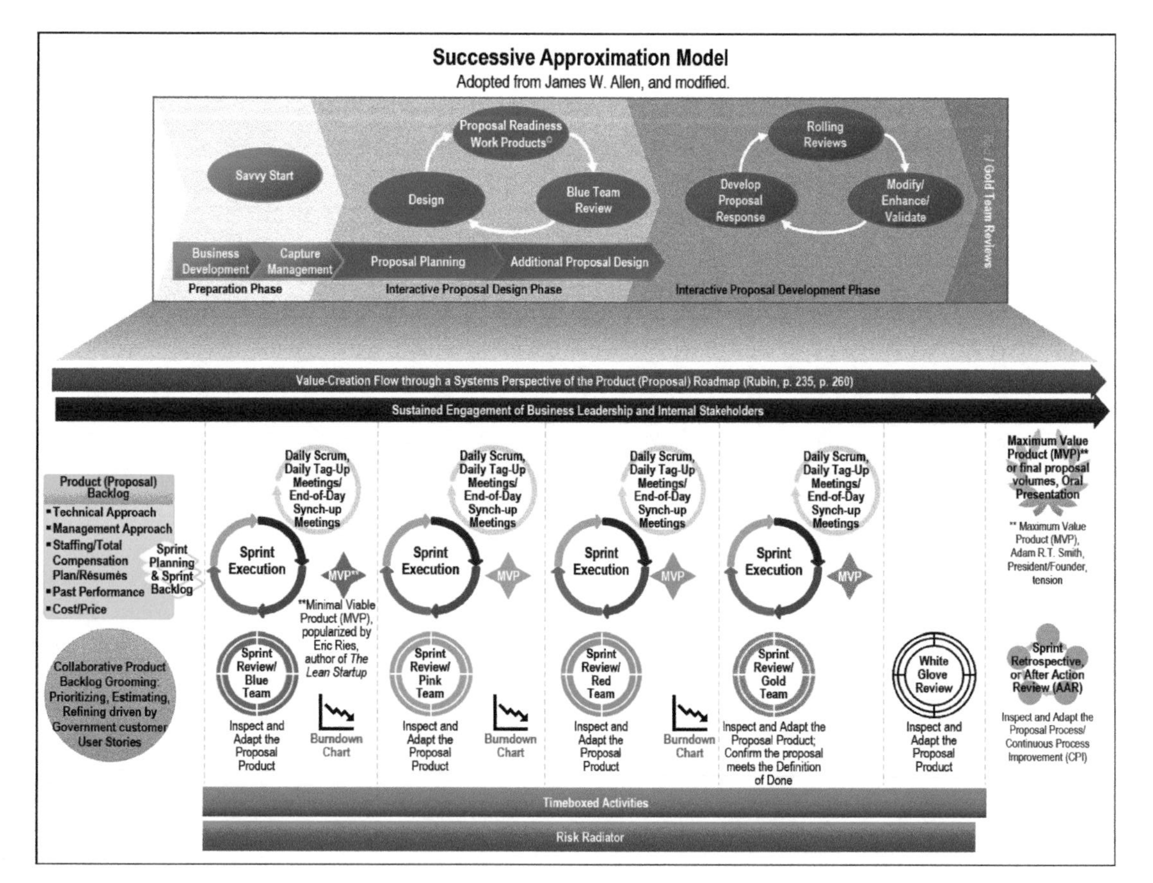

Figure 17.6 Looking at proposal development and review through the lens of SAM and Scrum.

interactive review, and ongoing enhancement are central, as opposed to only the traditional color review milestones, such as Pink, Red, and Gold Reviews.

Early, customer-facing business development efforts coupled with effective, solution-focused and strengths-centered capture management maximize our probability of winning, or P_{win}, in the public and the commercial sectors. In many instances, what follows next after the release of the final solicitation document or tender is far too focused on what I will call "proposal logistics"—schedule development, making writing assignments, page layout, and the like. These are important, to be sure, but should absolutely not be the primary drivers right out of the gate.

Organizations achieve better results—in effect, more contract awards—and spend fewer B&P dollars and cents when they do the planning, thinking, and initial illustrating before they do the writing. So the sequence should be PLAN-THINK-DRAW-WRITE. Foundational in the modified SAM is development and interactive review of Proposal Readiness Work Products, a term that I have coined and a process that I have copyrighted. Its origins lie in NASA's Technology Readiness Levels (TRLs). TRLs are a type of measurement system used to assess the maturity level of a particular technology (see Figure 17.7).

Proposal Readiness Work Products are analogous to TRLs in that the Proposal Readiness Work Products are assessed during the interactive Blue Team Review to determine their level of solution content and messaging maturity. Figures 17.8 and 17.9 present a populated Proposal Readiness Work Product for a NASA proposal.

The fundamental focus is on how a company's understanding informs its approach, as validated by relevant past performance. The approach, in turn, delivers benefits to the government or commercial-sector customer.

Proposal Readiness Work Products begin with an elevator speech. Think of getting into an elevator at Met Square in St. Louis. By the time you arrive at your chosen floor—along with your customer, you will want to have characterized the current technical environment associated with a given Statement of Work or Performance Work Statement section and introduced your company's high-level approach in this specific technical section. You also want to have begun to paint a clear picture of how your particular approach will deliver tangible

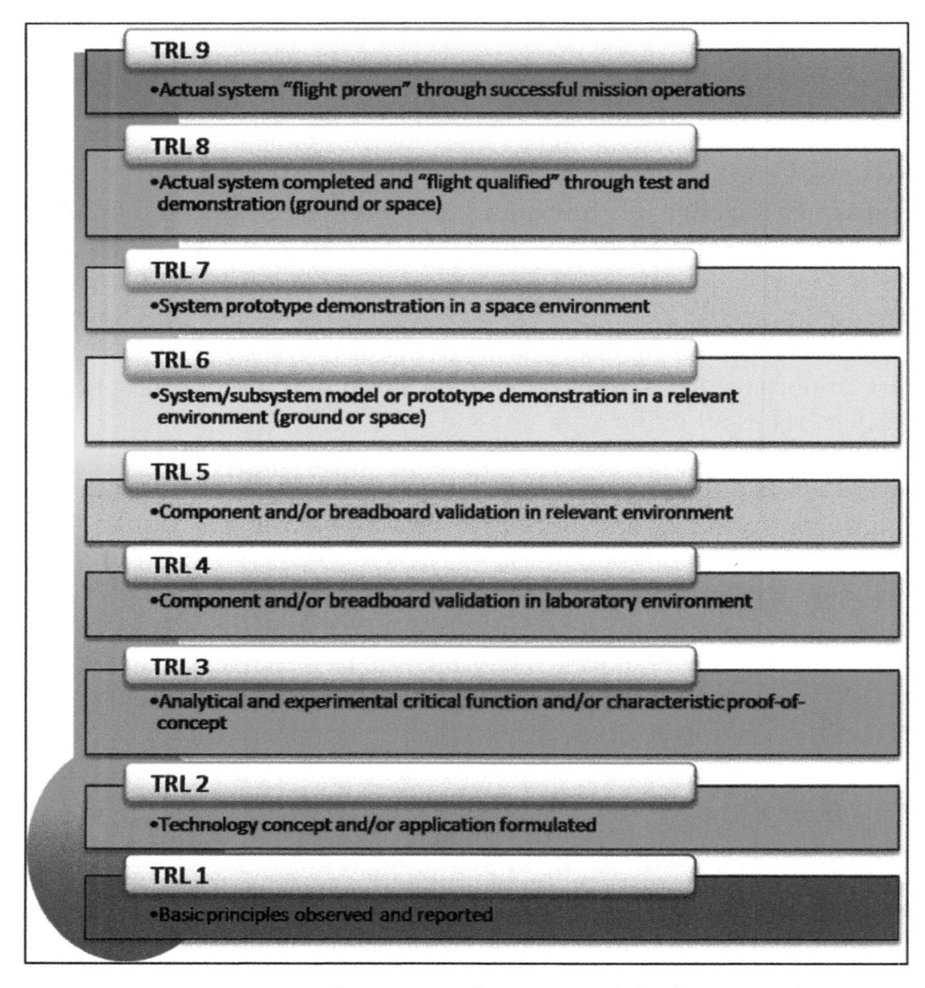

TRL 9
•Actual system "flight proven" through successful mission operations
TRL 8
•Actual system completed and "flight qualified" through test and demonstration (ground or space)
TRL 7
•System prototype demonstration in a space environment
TRL 6
•System/subsystem model or prototype demonstration in a relevant environment (ground or space)
TRL 5
•Component and/or breadboard validation in relevant environment
TRL 4
•Component and/or breadboard validation in laboratory environment
TRL 3
•Analytical and experimental critical function and/or characteristic proof-of-concept
TRL 2
•Technology concept and/or application formulated
TRL 1
•Basic principles observed and reported

Figure 17.7 NASA researcher, Stan Sadin, conceived the first TRL scale in 1974 with seven levels. In the 1990s, NASA adopted a scale with nine levels which gained widespread acceptance across industry and remains in use today. (*From:* [7].)

and intangible benefits that meet and actually exceed requirements of the federal government or commercial corporation—whomever your customer is—over the period of performance of the contract. And do this crisply and concisely—all in 6 to 8 sentences.

Also included as an integral part of the Proposal Readiness Work Product shown here is the T2 Chart, which I have copyrighted as well. The two T's stand for today and tomorrow. Today is the as-is

Approach to Innovations & Efficiencies		
Product, outcome, or deliverable	Change creation to process, products, or ideas	Cost savings improved product & performance
Proposal Support Services	Deliver seven, day-long proposal development seminars at no cost	Investment from [redacted] of more than $52,500
Management of Services	Realign Media Service management to reduce dual Managers from 2 WYEs to 1	Cost savings and avoidance; reduced management burden

Potential Risk	Mitigation
Budget constraints present the risk of staff reduction and loss of critical skills	Cross-train personnel; develop work processes to provide support for other tasks
Decreased usage of Media Services due to lack of end user awareness or higher cost	[redacted] uses integration and awareness methods to control costs and market to keep work "Inside the Gate"

Evidence of STRENGTHS

- Lean, Integrated Management results in continuous quality improvement and cost efficiency
- [redacted] improves contract optics and consolidates tracking and management of Pay for Services
- Best in Class Proposal development support at no cost, bringing money into GRC, "keep work inside the gates", supporting the communities of Northeast Ohio
- [redacted] includes a [redacted] incumbent contractor with excellent rating in Media Services for GRC

TA1.a.3.4 Business Process Support [C.8, CWA3.0, C.3.4]

Business Process Support (BPS) services meet the IT needs of [redacted] work areas and include support for applications, web databases, file storage, and servers. [redacted] approach to provide these services will leverage guidance from our **knowledgeable and experienced technical** staff, our **defined and robust approach to risk management**, a **focused QC program** strengthened by [redacted] methods, **industry leading IT tools**, and **proven technical processes**.

Understanding Opportunities

- LTID and the work/task areas that support [redacted] rely on BPS to provide critical IT services, including application development and server administration
- More than 50% of the work supported in BPS is in support of legacy/homegrown tools, applications, and databases across various disparate systems and platforms

Approach

- Skills: 3.4 – Business Process Management, Programming, Systems Analysis, Server Administration, Application Development, Technical Assistance, Enterprise Architecture, IT Systems Engineering, System Capacity Planning and Performance Management, IT Security, Planning and Integration
- Provide positive change, today *and sustained over time* , implementing new and innovative tools and technologies to improve efficiency and enhanced reporting
- Evaluate all custom applications in the first 90 days of the contract to identify efficiencies through consolidation, working with [redacted] to transition critical applications and infrastructure over time to standardized applications/platforms
- Adhere to an Agile industry approach to development, that defines and enforces quality standards, schedule, and documentation requirements in a prioritized fashion
- Provide tailored, near real-time contract tracking and reporting using iSite® while maintaining interface with Sharepoint

Past Performance Validation

- [redacted] applied an Agile Methodology on Forest Service Virtual Incident Procurement (VIPR) application development project that allowed users to begin to use a functional tool in a shorter period of time
- In support of NASA, [redacted] developed a highly praised, web-based Safety, Reporting and Tracking System (SRTS) that combined input from three databases
- [redacted] provides application development for full lifecycle software development and system architecture services in support of the NASA Bioastronautics Contract.

BENEFITS TO GRC LTID

- ★ Cost Effectiveness - reduced support hours for standardized tools
- ★ Efficiencies - streamlined operations, more useful programmatic and task order contract data, systems integration, ease of use and consistency across all business support elements, more clear and concise optics into operations
- ★ Meeting quality standards and schedules - includes stronger IT security measures, quicker application development, prioritization of IT needs
- ★ Reduced dependency on homegrown and legacy software

Figure 17.8 This is page 1 of the Work Product, which focuses on the elevator speech, understanding, approach, past performance validation, and benefits to the government.

Policy Alignment: CFR Title 14, Part 1260.28 & 1260.57, Title 35, Chapter 18; GLPD 2810.1; NASA-STD-8739; NPD 2091.1, 7500.2; NPR 2210.1, 7150.2, 7500.1; Public Law 96-480

Techniques & Rationale

will support the full spectrum of IT services from requirements definition, to design, engineering, installation, maintenance, sustainment, replacement, and disposal. We will look across the entire enterprise to identify commonalities and interdependencies among functional areas and exploit them to enhance performance and increase efficiency. In order to provide efficiencies and speed the time to delivery, we will evaluate all custom applications in the first 90 days of the contract to find efficiencies through consolidation — such as consolidating the many Bar Code applications and moving some applications to a web based, mobile-accessible platform to provide access in the field. Our team will suggest and implement new technologies and industry standard processes and applications without disruption to ongoing activities, improving the overall success of this critical function. For example, in support of new applications development, will leverages our proven Agile-based application development methodology, which will result in rapid development of the IT solutions, prioritizing needs and focusing first on those features that are most critical. We will also analyze existing legacy and home-grown applications and systems, and develop migration plans focused on retaining functionality while increasing the ease of support. Our approach helps identify and prioritize recommendations that will yield the greatest benefit to .

We will ensure all System Administrators have all the necessary skills and training needed to comply with personnel security investigation requirements as specified in NPR 2810.1 and will conduct classes in server consolidation methodologies, virtualization, and server management optimization techniques, enabling us to support the data center using fewer resources. will use commercial best practices to baseline all existing services, identify improvement opportunities, engineer solutions, and perform installations and upgrades. We will established a prioritized schedule of installations and manage the process from cradle to disposition.

Approach to Innovations & Efficiencies		
Product, outcome, or deliverable	Change creation to process, products, or ideas	Cost savings improved product & performance
Legacy Tools/ Applications Support	Identify COTS replacement	Low cost maintenance; enhanced functionality
Homegrown Apps/DBs on many platforms	Standardization of tools and platforms; transition to COTS	Direct operation and maintenance cost savings
Disparate systems	Integrating systems, leveraging shared data and functions	Enhanced optics; better integrated data reporting
Standard reporting	provides increased optics	Cost/schedule savings; Meaningful data

Potential Risk	Mitigation
Server down for extended period of time, exceeding performance requirements	Develop concrete Recovery/Restoration plans and backup solutions
Application corruption results in the loss of critical data	Develop Recovery/Restoration plan; implement modern applications/tool
Outdated application no longer supported by vendor	Identify replacements using risk management/business case analysis of our process
Lose staff resources who have special skills in supporting legacy system	Cross-train resources to eliminate single point failures; evaluate modernizing solutions

Evidence of STRENGTHS

- Agile Development, enhanced by targeted activities, will result in a rapid increase in functionality, at a dramatically reduced cost and time to implement
- Increased confidence in IT support, relying on industry-standard applications and databases, and reduced dependency on homegrown and legacy software

SUBFACTOR A: Technical Approach

Use or disclosure of data contained on this sheet is subject to the restriction on the title page of this proposal.

Figure 17.9 This is page 2 of the Work Product, which addresses innovations and efficiencies, potential risks and mitigation, and evidence of strengths.

environment—for example, the technical landscape, business objectives, and relevant mission goals—from the perspective of your customer. Tomorrow characterizes the to-be environment in terms of technical landscape, business objectives, and relevant mission goals. Again, this is the future from the perspective of your customer.

Together, the elevator speech and T2 Chart—along with 11 other key elements, make up one Proposal Readiness Work Product. Of particular value in terms of streamlining the proposal development process—when fully populated and then reviewed, vetted, and modified at the Blue Team Review, these Work Products become embedded directly into your proposal document. There is no traditional storyboard that requires further translation into proposal-ready form and format. In page-limited proposals, the Proposal Readiness Work Products conserve valuable space because they are presented as graphics, even 3-D figures.

Extraordinarily valuable as well is that when the Proposal Readiness Work Products are vetted and consensus is built regarding their content at the Blue Team, the proposal writers have a clear direction for developing narrative and associated additional graphics and tables. They are equipped with a robust 3-D map by which to navigate the Proposal Highway—think of it as Interstate 55 (from Louisiana to Illinois) or Interstate 70 (from Utah to Maryland).

So now we move to the interactive proposal development phase of the modified SAM's process. Here's where the model is very similar to Agile software development or APM. Leveraging the various Blue Team products, proposal writers then craft their specific sections and subsections. However, notice the critical difference: they do not wait for formal Red Team and other reviews to seek and obtain valuable feedback.

Instead, the capture lead establishes "rolling review" workflows—for example, from William—the primary author, on to Emma, a SME in the specific technology or process, and then on to Lucas, a senior program manager with direct experience with this specific customer.

More than 20 years ago, I had the opportunity to work closely with an outstanding professional capture manager. I would write a specific proposal section or subsection and place the hardcopy printout under his office door in the evening. He preferred paper copies so that he could use a red pen for markup. He was also a very-early-morning type of person. When I got to the office, he had already read, reviewed, commented on, and edited my work product. We continued in that manner to polish the proposal section until he determined that it was ready for the formal Red Team Review. He did this with other writers and content providers as well. The result was a much more mature and integrated document at the formal review stages such as the Red Team. Ongoing feedback was the pivotal difference-maker. This is strikingly superior to the traditional scenario of a group of proposal authors working largely in isolation from each other and the capture and proposal teams and then proposal management and production staff trying to "glue" the disparate pieces together for a formal review.

Better interim proposal sections and subsections translate into far more beneficial and value-added formal reviews. Then the focus can be on enhancing and further validating versus reworking major modules of text and graphics and/or rethinking the overall strategic direction for developing the proposal. Rolling reviews actually represent Lean Six Sigma practices in action. Reducing rework is essential for sustainable proposal operations. Rolling reviews are also analogous with "progressive elaboration," a term used by the Project Management Institute (PMI). Progressive elaboration is "the iterative process of increasing the level of details in a project management plan as greater amounts of information and more accurate estimates become available" [8].

So let's recap the modified SAM and its value to proposal development. This cyclical approach facilitates continuous improvement of proposal content and fine-tuning of messaging. There is time available during the majority of the proposal response life cycle to build in excellence along the way in bite-sized pieces. The modified SAM also supports the careful dovetailing of technical and management solutions and helps to ensure tight connections between résumés and past performance citations and the technical and management solutions being advanced. I cannot tell you how many times I review my customers' proposals in which the specific technical and management approaches being offered are not reflected in the content of résumés or in past performance write-ups.

ADDIE, SAM, Waterfall, and Agile

In the field of Instructional Systems Development, the conceptual model called *ADDIE*— Analysis, Design, Development, Implementation, and Evaluation—preceded the *SAM*. ADDIE is similar to the Waterfall Model, which is associated with the research and design approaches that Dr. Winston Royce created in 1970 [9]. Dr. Royce was an American computer scientist and director at the Lockheed Software Technology Center in Austin, Texas. Waterfall is a "formulaic approach to design that relied on each step being completed before moving on to the next" [9].

The U.S. Department of Defense (DoD) was a major advocate for Waterfall software development, which followed a "rigidly sequential, big design up front (BDUF), big test at the end, 'document everything' approach" [9, p. 1]. The *Agile movement* arose in part as a reaction to "analysis paralysis" associated with the waterfall software development methodology [10, p. 2].

Now we will turn our attention to the lower portion of Figure 17.6, with its depiction of the proposal development and review processes through the lens of APM. The most common application of APM is through Scrum. Agile projects are planned and executed in short iterations called sprints [11]. These cycles occur within timeboxes. The benefits of timeboxing encompass the following [12]:

- Establishes a work in progress (WIP) limit;
- Forces prioritization;
- Demonstrates progress;
- Motivates closure;
- Improves predictability;
- Avoids unnecessary perfectionism.

From a cross-cutting perspective, the APM process helps to create value through a systems perspective of the proposal roadmap. APM also promotes sustained engagement of your company's business leadership and internal stakeholders throughout the proposal development life cycle. The entire capture and proposal team is kept aware through a risk radiator and dependency board.

As discussed by O'hEocha and Conboy [13], user stories are a common practice in Agile methods for feeding user requirements into the software or proposal development process. Unlike traditional requirements engineering approaches, they do not call for comprehensive specification of the solution up-front. User stories express user-centric functionality, and are written in a story style. They reflect what the user would like the system (read, proposal solution) to do, rather than how it should do it. Along the way through the process depicted in Figure 17.6, there are burndown chart icons. Certified Scrum Master Maryann Lesnick [14] suggested using burndown charts to "improve time management and increase accountability" and visibility.

Under APM, the entire goal of the proposal development and review process is one of crafting a maximum value product (MVP), a term that originated with Adam R. T. Smith, president and founder of tension. Throughout the 30 to 45-day Scrum, at each Sprint Review (e.g., Blue Team, Pink Team), the interim target is to develop a minimal viable product. This term was popularized by Eric Ries, the author of *The Lean Startup* [15]. Together, daily tag-up calls, end-of-day sync-up calls, and Sprint Reviews provide a high level of transparency into proposal progress and provide the forum in which to enhance proposal quality [16].

Scrum comprises three key roles [11], which together make up the Scrum team (capture team/proposal team) (see Figure 17.10).

For large, complex RFP responses with multiple development teams, the Scrum of Scrums (SoS) methodology is applicable. SoS was first implemented in 1996 by Jeff Sutherland and Ken Schwaber.

An AAR, or "retrospective" in Agile parlance, stands as an important proposal process assessment that is an integral part of Continuous Process Improvement (CPI). Conversely, the Sprint Reviews constitute proposal product assessments.

17.4 AGILE IN CONTEXT

It turns out that Agile processes are not all that new. Dr. Walter A. Shewhart, an American physicist, engineer, and statistician, began improving products and processes through iterative cycles in the

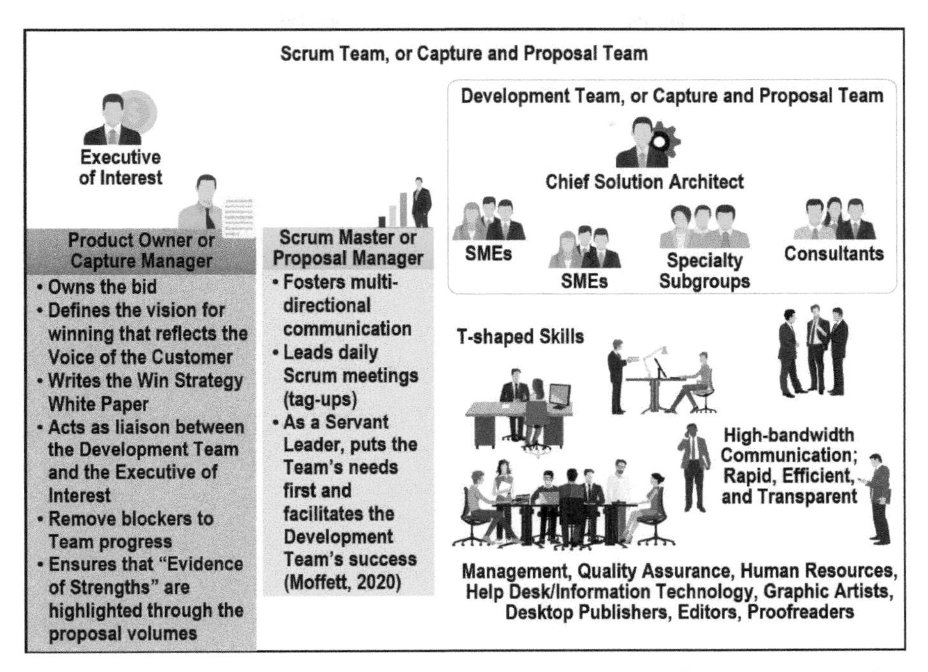

Figure 17.10 With B&P dollars infused by the Executive of Interest, the product owner (capture manager), together with the Scrum master (proposal manager), works closely with the development team (capture and proposal team) to build an MVP.

1930s [17]. This practice was later modified by W. Edwards Deming to become the Plan-Do-Study-Act (PDSA), also known as Plan-Do-Check-Act (PDCA), a cycle for continuous improvement and quality management. Up through the 1980s, the U.S. military, NASA, IBM, Honda, Toyota, Canon, and others continued to experiment with and evolve concepts and practices that we recognize today as Agile. These ideas led to the publication of the Agile Manifesto in 2001 and identification of the common values and principles for improving the approach to system development projects. Currently, several varieties of Agile-based methodologies are used in these efforts, including Scrum, Extreme Programming (XP), and in some cases, Kanban [17].

Selected Bibliography

Angularjs, D., "Agile Proposal Development: A Spotlight on Scrum," July 19, 2020, https://www.proposalreflections.com/2019/02/agile-proposal-development-spotlight-on.html.

Evans, L., "The ADDIE Model for Instructional Design [+Pros/Cons & FAQs]," University of San Diego Online, 2023, https://onlinedegrees.sandiego.edu/addie-model-instructional-design/.

Herrholtz, K., "Rapid Instructional Design with SAM," *eLearning Industry,* March 26, 2020, https://elearningindustry.com/tips-implement-sam-model-in-elearning.

Kayes, A., "A Guide to Agile Proposal Development and Management," February 3, 2022, https://info.ksiadvantage.com/blog/a-guide-to-agile-proposal-development-and-management.

Moffett, F., "The Technical Proposal Manager as ScrumMaster," *Winning the Business,* October 20, 2020, https://winningthebusiness.com/the-technical-proposal-manager-as-scrummaster/.

Wintarti, A., et al., "The Instructional Design of Blended Learning on Differential Calculus Using Successive Approximation Model," *Journal of Physics: Conference Series,* 2019, p. 1417.

References

[1] U.S. Department of Defense, *Source Selection Procedures: Defense Federal Acquisition Regulation Supplement Procedures, Guidance, and Information,* Subpart 215.3—Source Selection, August 20, 2022, p. 23.

[2] Thalheim, B., "The Theory of Conceptual Models, the Theory of Conceptual Modelling and Foundations of Conceptual Modelling," in D. W. Embley, and B. Thalheim (eds.), *Handbook of Conceptual Modeling: Theory, Practice, and Research Challenges,* New York: Springer, 2011, p. 5.

[3] Jung, H., H. Lee, and Y. Kim, "Advanced Instructional Design for Successive E-Learning: Based on the Successive Approximation Model (SAM)," *International Journal on E-Learning,* Vol. 18, No. 2, 2019, p. 191, https://www.researchgate.net/publication/331829482_Advanced_Instructional_Design_for_Successive_E-Learning_Based_on_the_Successive_Approximation_Model_SAM.

[4] Ali, C. A., S. Acquah, and K. Esia-Donkoh, "A Comparative Study of SAM and ADDIE Models in Simulating STEM Instruction," *African Educational Research Journal,* Vol. 9, No. 4, 2021, p. 845, https://files.eric.ed.gov/fulltext/EJ1324100.pdf.

[5] Ciric, D., et al., "Agile Project Management Beyond Software Development: Challenges and Enablers," Conference Paper, 2021, p. 245, https://www.researchgate.net/publication/351811492_Agile_Project_Management_beyond_Software_Development_Challenges_and_Enablers.

[6] Allen, M. W., *Leaving ADDIE for SAM: An Agile Model for Developing the Best Learning Experiences,* American Society for Training & Development (ASTD) Press, 2012.

[7] https://www.nasa.gov/directorates/heo/scan/engineering/technology/technology_readiness_level.

[8] Lotffy, E. M., "Progressive Elaboration of Project Management Processes," *PM World Journal, V(V),* 2016, p. 8.

[9] Eller, R., "Implementing Design into Instruction: Intersections Between the Waterfall Model and ADDIE Model," *Assumption University-eJournal of Interdisciplinary Research (AU-eJIR),* Vol. 1, No. 1, 2015, p. 65.

[10] Palmquist, M. S., "Parallel Worlds: Agile and Waterfall Differences and Similarities," Carnegie-Mellon University, Technical Note CMU/SEI-2013-TN-021, 2013.

[11] Kayes, A., "Agile Proposal Development: A Spotlight on Scrum," February 4, 2019, https://www.proposalreflections.com/2019/02/agile-proposal-development-spotlight-on.html.

[12] Rubin, K. S., *Essential Scrum: A Practical Guide to the Most Agile Process,* Reading, MA: Addison-Wesley, 2013, p. 63.

[13] O'hEocha, C., and K. Conboy, "The Role of the User Story Agile Practice in Innovation. *Lecture Notes in Business Information Processing,* Vol. 65, 2010, pp. 20–30.

[14] Lesnick, M., "Can Your Proposal Process Be More Agile?" *2014 Mid-Atlantic Conference & Expo,* Tysons Corner, VA, November 14, 2014, https://slideplayer.com/slide/4430147/, p. 21.

[15] Ries, E., *The Lean Startup: How Today's Entrepreneurs Use Continuous Innovation to Create Radically Successful Businesses,* New York: Crown Publishing/Random House, 2011.

[16] Chowdhury, M., "Implementing Agile in Bid and Proposals," LinkedIn, September 3, 2019, https://www.linkedin.com/pulse/implementing-agile-bid-proposals-chowdhury-cp-apmp-sa-csm-pmp/.

[17] California Project Management Office (CA-PMO)/California Department of Technology, *Understanding Agile,* 2017, p. 13.

18

BUILDING THE COST/PRICE VOLUME[1]

Among the keys to building compliant and robust cost/price proposals (see Figure 18.1) are having a fundamental understanding of pricing terminology that appears in the FAR, Cost Accounting Standards (CAS), and the role of the Defense Contract Audit Agency (DCAA),[2] as well as other cognizant auditors.

18.1 OVERVIEW OF THE FAR

Per the General Services Administration,[3] "The Federal Acquisition Regulation (FAR) is the primary regulation for use by all executive agencies in their acquisition of supplies and services with appropriated funds. The FAR also contains standard solicitation provisions and contract clauses and the various agency FAR supplements." There are

1. The content for this chapter was developed by Kevin M. McQuade, president and CEO of MSM Group, Inc. (www.msmgroupinc.com), and Matt McKelvey, president of The McKelvey Group (TMG) (https://themckelveygroup.com).
2. The DCAA is the largest and most recognized cognizant auditor in the federal government. Their guidelines and procedures are followed by auditors from other agencies. Therefore, for the remainder of this chapter, any reference to the DCAA by default means any cognizant government auditor performing audit and financial advisory services for acquisition and contract administration.
3. This quotation comes from https://www.gsa.gov/policy-regulations/regulations/federal-acquisition-regulation.

Figure 18.1 U.S. Mint Proof Set and two $2 bills. (Photograph © Dr. R. S. Frey.)

some organizations within the federal government exempt from the FAR,[4] but they are very much the exception. In addition to the FAR, each executive agency has a supplemental FAR that provides additional regulations to perform work for the specific agency. The acquisition process is the process through which the government purchases ("acquires") goods and services. It does not regulate the purchasing activities of private-sector firms, except to the extent that those parts of it are incorporated into government solicitations and contracts by reference. The purpose of the FAR is to provide uniform policies and procedures for acquisition. Among its guiding principles are to have an acquisition system that satisfies customers' needs in terms of cost, quality, and timeliness; to minimize administrative operating costs; to conduct business with integrity, fairness, and openness; and to fulfill other public policy objectives. As a starting point, let's ensure that we have factual understanding of important pricing terms and concepts.

18.2 DEFINING KEY PRICING TERMINOLOGY

18.2.1 Cost Objective

As defined in FAR 31.001, a cost objective is a function, organizational subdivision, contract, or other work unit for which cost data are

4. Organizations such as the Central Intelligence Agency (CIA), United States Postal Service (USPS), U.S. Senate, and Tennessee Valley Authority (TVA), as well as specific procurement types such as Other Transaction Authority (OTA), have their own acquisition regulations and are not subject to the FAR.

desired and for which provision is made to accumulate and measure the cost of such elements as processes, products, jobs, and capitalized projects.

18.2.2 Final Cost Objective

A final cost objective, as defined in FAR 31.001, is a cost objective that has allocated to it both direct and indirect costs and, in the contractor's accumulation system, is one of the final accumulation points. Cost accumulation refers to the manner in which costs are collected and identified with specific customers, jobs, orders, departments, and processes.

18.2.3 Direct Costs

As defined in FAR 31.202, direct costs are identifiable to a final cost objective (a particular contract) such as direct material and direct labor. All costs identified specifically with a given contract are direct costs for that contract and cannot legally be charged to another contract either directly or indirectly. No cost can be charged to a contract as a direct cost if other costs incurred for the same purpose in like circumstances have been charged as an indirect cost.[5] The most important principle here is the consistent definition and charging of direct costs across all contracts.

18.2.4 Direct Labor

Direct labor (DL) is the cost of all the time (hours) that W2 employees spend on a particular final cost objective that is directly related to the contract objectives. Direct labor does not include the labor performed by subcontractors or consultants. Their time is charged directly but separately from direct labor. The direct labor cost is the number of hours spent by an individual on the job times the dollars per hour which that person is paid. Direct labor is the salary and wages of employees performing work identifiable to a final cost objective, such as a contract. It does not include the salary and wages of employees for time spent for paid leave (when recovered in fringe benefits), overhead, general and administrative (G&A), or unallowable activities.

5. Some companies charge certain costs directly (such as PMO or facilities), while others charge them indirectly.

18.2.5 Subcontractor Labor

Subcontractor labor is the cost of all the time (hours) that subcontractors and consultants spend on a particular final cost objective that is directly related to the contract objectives. While the cost is based on hours, the charge to the prime contractor may be hourly or as part of a fixed price. Regardless, subcontractors and consultants must track their time in their accounting systems and provide this information to the government upon request.

18.2.6 Labor Categories

Labor categories (LCATs) and their descriptions identify the skills, knowledge, abilities, experience, training, and educational requirements of the labor to be provided on a contract. Contractors use these descriptions in developing their prices/costs for labor. Labor categories are used by both W2 employees and subcontractor/consultant labor.

18.2.7 Other Direct Costs

Other direct costs (ODCs) are cost accumulation elements (other than labor and material) in the contractor's accounting system that can be specifically assigned to a contract or other final cost objective. Other direct costs are other costs charged directly to the government that have not been included in proposed material, direct labor, indirect costs, or any other category of cost. ODCs can include travel, expenses, computer time, special tooling, relocation expenses, preproduction and start-up costs, packaging and transportation costs, royalties, spoilage and rework, federal excise taxes, and photoreproduction costs, along with subcontracts and services.

18.2.8 Indirect Costs

As defined in FAR 31.203, indirect costs are not directly identified with a single, final cost objective, but identified with two or more final cost objectives. After direct costs have been determined and charged directly to the contract or other work, indirect costs are those remaining to be allocated to intermediate or two or more final cost objectives. No final cost objective can have any cost allocated to it as an indirect cost if other costs incurred for the same purpose in like circumstances have been included as a direct cost of that or any other final cost objective.

The FAR does not define the indirect rates contractors must use or the specific costs charged to each indirect pool. However, there are common indirect rates used across government contractors. These typical basic indirect rate structure elements include fringe benefits, overhead, procurement, and G&A.

18.2.9 Fringe Benefits

In a separate indirect pool,[6] most government contractors accumulate the costs associated with three main components of benefits afforded to their employees, namely: (1) paid leave, (2) employer-paid payroll-related taxes, and (3) employer-paid benefits. Absences include such paid time off as holiday, vacation, and sick leave. Some contractors combine vacation and sick leave into paid time off (PTO). Other examples of paid leave include military, family, jury duty, and bereavement. Payroll-related taxes encompass statutory elements such as the Federal Insurance Contributions Act (FICA), the Federal Unemployment Tax Act (FUTA), and the State Unemployment Tax Act (SUTA). Employee benefits include medical, retirement, life and disability insurance, tuition, health savings accounts, and other employee-specific benefits.

The allocation base for fringe benefits is W2 labor dollars.

18.2.10 Overhead

Overhead (OH) includes those costs not directly related to a final cost objective. These consist of indirect costs that are associated with the production of goods and services and that benefit more than one contract or cost objective. Examples of overhead costs can include task management and reporting systems, certification training, professional conferences, and recruiting fees. There are no prescribed overhead costs per the FAR, so the key is for a contractor to consistently charge costs here rather than directly. A contractor cannot charge similar costs sometimes directly but other times add them to OH.

6. An indirect cost pool is a logical grouping of indirect costs with a similar relationship to the cost objectives. For example, engineering overhead pools include indirect costs that are associated with engineering efforts.

Note that some companies combine overhead and fringe benefits into a single indirect rate. This is not as common today as it was from the 1980s to the early 2000s.

The allocation base for labor overhead is commonly direct labor plus direct fringe benefits. Some contractors may have the base as simply direct labor. For manufacturing overhead, the base is defined by the company.

18.2.11 Procurement

Procurement costs are those costs related to managing the purchases of goods and services that do not require "value added" by the contractor. Commonly called "pass-through" costs, these direct costs are necessary to the performance on a contract but do not require substantial indirect management by the contractor. Therefore, the costs associated with acquiring and delivering materials and subcontracted labor are often charged to a procurement rate. There are no prescribed procurement costs per the FAR, so the key is for a contractor to establish if they have significant pass-through costs. For most service contractors, this would mean a subcontractor burden rate. For manufacturers and resellers, this would mean a material burden rate. For larger contractors, it may be a combined material and subcontractor rate. There is no requirement to have this rate per the FAR. It is usually driven when there is a significant amount of pass-through and the establishment of a procurement rate makes logical sense. The procurement costs typically include purchasing, contracts, and associated facilities or system costs.

18.2.12 G&A

G&A expenses represent the cost of activities that are necessary to the overall corporate operation of the business as a whole, but for which a direct relationship to any particular cost objective cannot be shown. Indirect G&A costs normally include executive, human resources, accounting, legal, internal information technology, bid and proposal, business development, corporate facilities, payroll administration, depreciation, and office expenses. There are no prescribed G&A costs per the FAR, so the key is for a contractor to consistently charge costs here rather than directly. A contractor cannot charge similar costs sometimes to G&A while other times directly or to other indirect pools.

The allocation base for G&A is commonly either total cost input (TCI), modified total direct cost (MTDC), and value added.

18.2.13 Fully Burdened Rate

A contractor's fully burdened rate (FBR) is the total price for delivering an employee to perform work for the government. The FBR is calculated by applying all appropriate indirect rates to the employee's direct labor rate or a subcontractor/consultant's bill rate. A fully burdened rate can be specific for an individual or an average across multiple personnel within a labor category. The factor applied to direct and subcontractor/consultant labor to arrive at the total cost of labor is often called a wrap rate. For example, if a contractor has an employee with a direct labor rate of $100.00 and a wrap rate including fee of $1.65, then the total price per labor hour would be $16.50. The wrap rate is the applicable total multiplier (fringe, overhead, procurement, G&A, and fee) that is applied to an individual's or labor category hourly rate to the fully burdened rate for an invoice in a time and material (T&M) or cost plus contract.

18.2.14 Fee

Both the government and contractors recognize that profit (fee) is a motivator of efficient and effective contract performance. Negotiations aimed merely at reducing prices by reducing profit, without proper recognition of the function of profit, are not in the government's interest. When performing a profit analysis, the government will take into consideration factors such as performance risk, contract type risk, facilities capital employed, and cost-efficiency.

18.2.15 Unallowable Costs

An unallowable cost is a perfectly legal cost that the federal government will not pay for directly or indirectly. Unallowable costs are discussed in FAR 31.203-6. A contractor may also choose to voluntarily identify other costs as unallowable. Costs that are expressly or voluntarily unallowable must be identified and excluded from any billing, claim, or proposal applicable to a government contract. These unallowable costs also include directly associated costs. A directly associated cost is any cost that is generated solely as a result of incurring an unallowable cost. When an unallowable cost is incurred, its directly associated costs are also unallowable. Examples of FAR-defined unallowable costs include bad debts, alcohol, charitable contributions or donations, interest, and entertainment.

18.3 OVERVIEW OF FEDERAL TRAVEL REGULATIONS

FAR 31.205-46 states that costs for transportation, lodging, meals, and incidental expenses incurred on official company business are allowable contract costs provided that the method used results in a "reasonable" charge. These travel costs will be considered reasonable as long as they do not exceed the maximum per diem rates in effect as set forth in the Federal Travel Regulations (FTR) or the Joint Travel Regulations (JTR) for travel outside of the contiguous United States. The FTR and JTR policies and procedures ensure accountability of taxpayers' money. FTR is contained in the Code of Federal Regulations (CFR) Chapters 300–304. The purpose of the FTR is to interpret policy requirements to ensure that official travel is conducted in a responsible, most cost-effective manner. It also strives to communicate the policies in a clear manner to federal agencies and employees.

18.4 OVERVIEW OF THE COST ACCOUNTING STANDARDS

The Cost Accounting Standards (CAS) are a set of 19 standards and rules promulgated by the U.S. government for use in determining costs on negotiated procurements. The purpose of the CAS is to achieve consistency and uniformity in cost accounting principles. These principles must be followed by contractors in the several phases of contract pricing, including estimating accumulating and reporting of costs. The CAS are similar in purpose to the generally accepted accounting practices (GAAP) in the commercial industry, but are absolute for CAS-covered contracting.

There are three levels of CAS coverage.

- *Exempt:* This contractor is exempt from the CAS. All small businesses are considered exempt until they trigger the next level of coverage.

- *Modified:* This is triggered when a contractor receives an award of a $7.5 million CAS-covered contract. Modified coverage requires the application of CAS 401, 402, 405, and 406.
- *Full:* This is triggered when a contractor receives an award of a $50.0 million CAS-contract or cumulative awards of $50.0 million in CAS-covered contracts in the past year. Full coverage requires the application of all CAS.

Note that the threshold can change and therefore contractors should stay aware of the current thresholds.

18.5 OVERVIEW OF RELEVANT GAAP

The GAAP refer to the common set of accounting concepts, standards, and procedures that represent a general guide. GAAP principles are those that have substantial authoritative support or are based on accounting practices accepted over time by prevalent use by the Financial Accounting Standards Board (FASB), American Institute of Certified Public Accountants (AICPA), and Accounting Principles Board (APB), among others. The end products of the accounting cycle, the financial statements (e.g., balance sheet and income statement), are prepared in accordance with GAAP. Where CAS is silent, GAAP applies.

18.6 ADEQUACY OF THE CONTRACTOR'S INTERNAL ACCOUNTING SYSTEM

FAR Part 31 defines the requirements for an adequate accounting system. FAR 16.301-3 requires that a contractor's accounting system be adequate for determining costs applicable to the contract prior to the award of a cost-reimbursable contract, grant, Small Business Innovation Research (SBIR), or Broad Agency Announcement (BAA). An adequate accounting system is not an evaluation criterion. It is a basic contract requirement with a pass/fail determination. The FAR states that a contract vehicle may only be awarded to an offeror who is determined to have an adequate accounting system by the Defense Contract Audit Agency (DCAA).

So, what is an adequate accounting system? An adequate accounting system is one that conforms with GAAP, produces equitable results and is verifiable, is applicable to the contemplated contract(s), and is capable of being followed consistently. The DCAA may perform a preaward accounting survey to determine accounting system adequacy. A preaward survey is an evaluation of a prospective contractor's ability to perform a proposed contract. In order to be determined to be adequate, your accounting system must have:

1. Proper segregation of direct costs from indirect costs;
2. Identification and accumulation of direct costs by contract;
3. A logical and consistent method for allocation of indirect costs to intermediate and final cost objectives;
4. Accumulation of costs under general ledger control;
5. A timekeeping system that identifies employees' labor by intermediate or final cost objectives and a labor distribution system that charges direct and indirect labor to the appropriate cost objectives;
6. Interim (at least monthly) determination of costs charged to a contract through routine posting to books of account;
7. Exclusion from costs charged to government contracts of amounts that are not allowable pursuant to FAR Part 31, Contract Cost Principles and Procedures, or other contract provisions;
8. Identification of costs by contract line item and units if required by the proposed contract;
9. Segregation of preproduction costs from production costs.

Note that an accounting system is not just the accounting software used. An accounting system is composed of the software, written policies and procedures, training of personnel, and internal controls.

18.7 THE DCAA

Organizations that wish to provide goods and services to the federal government must be prepared to comply with regulations promulgated by a host of government agencies. The DCAA is the most prominent audit and financial advisory services agency for acquisition and contract administration in the government contracting arena. The DCAA provides standardized contract audit services for the DoD, as well as accounting and financial advisory services regarding contracts and subcontracts responsible for procurement and contract administration. These services are provided in connection with negotiation, administration, and settlement of contracts and subcontracts. The DCAA also provides contract audit services to some other government agencies. Importantly, the DCAA follows generally acceptable government auditing standards (GAGAS) when conducting its audits. Finally, cognizant auditors in other agencies primarily follow the guidelines and policies of the DCAA.

The DCAA's major areas of emphasis include:

1. Internal control systems;
2. Management policies;
3. Accuracy and reasonableness of cost representations;
4. Adequacy and reliability of records and accounting systems;
5. Financial capability;
6. Contractor compliance with contractual provisions having accounting or financial significance such as the Cost Principles (FAR Part 31), the "Cost Accounting Standards Clause" (FAR 52.230-2), and the clauses pertaining to the Truth in Negotiations Act (TINA) (FAR 52.215-10, -11, -12, and -13).

The DCAA stands as a distinct agency within the DoD and reports directly to the Under Secretary of Defense (Comptroller). The agency is organized into five regions and a field detachment and has individual field audit offices (FAOs) and suboffices throughout the United States and overseas. The DCAA's main service is performing contract audits requested by representatives of various military and civilian acquisition organizations and by the Defense Contract Management Agency (DCMA). Each year, the DCAA audits thousands of contractors, using a risk-based approach to identify the highest priority audits. At the completion of an audit, an audit report is furnished to the requestor for use in negotiations and/or in determining current or future contractor costs. An audit report can also be used to determine compliance with existing regulations and contractual requirements.

18.8 RFP REVIEW

Production of the cost and price volume of the proposal is an exercise that parallels the approach incorporated for the rest of the proposal response.

First, read the solicitation—all of it (even the boring parts). Then reread the solicitation—all of it—*again*. Section L instructs offerors how to respond and contains specific language pertaining to proposal preparation and adherence, provides an index of proposal contents and structure, identifies cost proposal evaluation factors, and is intended to correlate to Section M on a one-to-one basis (note, however, that this is sometimes not the case). However, do not neglect Sections B to K and any attachments in your review. Information that affects pricing is spread throughout the RFP and if you limit your review to just Sections L and M, you will miss critical information. If you have any questions, ask before the question-and-answer period for the given solicitation closes; do not assume that you have the correct understanding of the RFP instructions. Have someone (not the preparer) check the proposal against the solicitation requirements.

18.9 IMPORTANCE OF ALIGNING THE COST PROPOSAL WITH THE TECHNICAL AND MANAGEMENT PROPOSALS

Cost proposals should absolutely not be produced in a vacuum. While the technical, management, past-performance, cost, and/or any other volumes are separate factors in the evaluation, the proposed price and underlying rates may be considered by the government to be significant indicator of the offeror's understanding and ability to perform the statement of work, the statement of objectives, and other requirements of the RFP. It is therefore extremely important that the cost information be consistent with the technical and management approach. The programmatic schedule to perform the technical work must align precisely with the schedule used in developing cost information. The staffing levels must be consistent and the labor costs must link directly with the skill levels discussed in the technical proposal. The total price estimated must include the costs (labor or other) of all activities described in the technical approach. Likewise, the approach articulated in the other volumes should also be clearly evident in the cost volume. Strategies and approaches for lines of authority, span of control, total compensation, and project management office (PMO) responsibilities all must align with the pricing approach and resultant final cost volume methodologies. Because all portions of your response to an RFP will be used to evaluate your company's understanding of the services and products required, the government may adjust your overall technical score based upon the degree of realism in price.

18.10 COST ESTIMATING METHODS

There are a number of important techniques that may be applied to estimate your proposed price, as shown in Table 18.1.

18.11 METHODOLOGY FOR DETERMINING SALARY RANGES

There are many potential sources for salary information. These can include actual personnel, representative personnel, survey data, market data, technical/hiring manager experience, wage determinations, and sometimes RFP-provided information. The key to determining reasonable and realistic salary data when actual salaries are not available is to find at least three sources to compare and eliminate outliers.

Reputable sources for salary survey data include the Economic Research Institute (ERI), Western Management Group Government Contractors Compensation Survey (WMG), and Willis Towers Watson Data Services (WTW).

18.12 METHODOLOGY FOR COMPUTING LABOR ESCALATION

Providing local market-informed competitive compensation to employees is a critical component in retaining the current contract workforce. Direct labor escalation in the cost proposal is the year-to-year increase in the direct labor base resulting from merit increases, cost-of-living trends, movement within the company, and changes in the size and composition of the workforce. Direct labor escalation factors are often based on a number of sources including internal historical data, the Bureau of Labor Statistics (BLS), commercial databases such as IHS Global Insight, and local salary survey data. When the correlation of labor rate data between multiple sources is strong and

Table 18.1
Spectrum of Cost-Estimating Methodologies

Estimating Method	Characteristics
Analogy	Estimate of costs based on historical data of a similar item. The analogy method compares a new or proposed system with one that is an analogous (i.e., similar) system and typically acquired in the recent past for which there is accurate cost and technical data. There must be a reasonable correlation between the proposed system and the historical system. The estimator makes a subjective evaluation of the differences between the new system of interest and the historical system. The analogy method is typically performed early in the cost-estimating process, such as the pre-Milestone A and Milestone A stages of a program.
Engineering (bottom-up)	The most detailed of all the techniques and the costliest to implement. It reflects a detailed buildup of labor, material, and overhead costs.
Parametric (top-down)	A cost-estimating methodology using statistical relationships between historical costs and other program variables such as system physical or performance characteristics, contractor output measures, or staff power loading.
Actual costs	The actual cost method uses the actual cost of the previous production lot adjusted for inflation, labor efficiencies, material cost, technology changes, and other factors. An actual cost is a cost sustained on the basis of costs actually incurred and recorded in accomplishing the work performed within a given time period, as distinguished from forecasted or estimated costs.

confidence level is high, the minimum and maximum salary ranges proposed will support the overall proposed price as well as provide documentation for the audit file.

18.13 HIGHLIGHTS OF THE PROCESS TO CALCULATE INDIRECT COSTS

The development of indirect rates and resultant costs in pricing a proposal is a major step in the overall pricing process. The cost accounting structure for indirect rates should be suited to match a given business unit's operations, and the allocation base should be the best representation of the causal/beneficial relationship between costs and cost objectives. Indirect rate cost accounting structure should be designed for maximum cost recoverability and competitiveness.

A well-written set of policies and procedures should be in place in your company that documents estimating procedures; the DCAA auditors will look for consistency among estimating, accumulating, and reporting practices. Estimating indirect costs to be applied to a particular proposal starts with an understanding of the pools, bases, and resultant indirect rate. Pools are defined by placing costs in logical groupings (e.g., fringe costs, overhead costs, G&A costs, and unallowable costs). The allocation base is defined as some measure of the cost that can be used to allocate the pool cost (e.g., direct labor hours, direct labor dollars, and machine hours).

18.14 DETERMINING ANNUAL PRODUCTIVE LABOR HOURS

Direct labor is estimated on the basis of productive effort. Productive effort is the estimated number of hours required to perform the work. Vacations, holidays, sick leave, and any other paid absences are not to be cited as direct labor and must be identified separately and priced or included in indirect cost. Productive hours may be called direct productive labor hours (DPLH) or something similar.

It is important to distinguish between DPLH and pay hours. Pay hours for salaried employees normally is calculated as 40 hours a week for 52 weeks, or 2,080 hours per year. Dividing an employee's salary by 2,080 hours will give you the direct labor rate. A DPLH is always lower than the pay hours. For example, an employee with 10 holidays per year, 2 weeks of vacation, and 1 week of sick would have a DPLH of 1,840. This is calculated as 2,080 – 80 hours of holiday – 80 hours of vacation – 40 hours of sick time. This represents the remaining time available to work after paid leave.

18.15 METHODOLOGY TO COMPUTE FEE

In determining the level of fee to propose, contractors need to consider a number of factors as described in FAR 15.404-1. These include contractor incentive, complexity of work, management effort, cost risk, capital investment, financing of operations, and past performance.

18.16 COST VOLUME NARRATIVE AND PRODUCTION

In addition to responding to the requirements by including the detailed pricing in the proposal, the cost volume should include a narrative section that is compliant with the requirements of the RFP. The narrative should follow the requirements in the RFP. However, in general, the narrative should articulate all pricing and estimating techniques in detail, including projections, rates, ratios, percentages, and factors, as required. You should also provide a complete disclosure of the methods you used in determining the classifications of cost (direct versus indirect). Many contractors also furnish a synopsis of their accounting policies and procedures. The proposed indirect rates should be supported by source, forecasts, dollar values, factors, and substantiating rationale for each rate. A well-written, comprehensive, and logically connected narrative demonstrates competence and increases overall win probability. A competitive proposal is one that is compliant, credible, compelling, consistent, and cost-effective. Before submitting the proposal, be sure to arrange for an independent edit of the final text, recheck the final against RFP requirements, include a compliance matrix in the submittal, prepare a cover letter for signature, refine the final graphics, test the print final version (even if delivering electronically), and arrange for delivery in exact accordance with the RFP instructions.

18.17 AUDIT FILES

Your company needs to build and maintain a complete, separate audit file during the proposal process. This will ensure that the DCAA auditor has all information necessary to conduct a comprehensive audit of the proposal. It will also provide documentation to support subsequent revisions and operations after contract award. Supporting information encompasses:

- Evidence of consistency between your company's estimating and accounting systems;
- Documentation of cost presentation by cost element at the summary level;
- Subsidiary schedules that support summary-level costs;
- Adequate documentation readily traceable to estimates in detailed schedules and to summary level;
- Written rationale defining the basis for costs;
- Validation for basis of estimate requirements;
- History of prior bid proposal audits;

- Schedules that are annotated with references to detailed support;
- Written estimating system methods/procedures.

18.18 SPECIAL TOPICS

18.18.1 Uncompensated Overtime

Uncompensated overtime is the unpaid hours worked over and above the standard 40 hours for exempt salaried employees. The DCAA is concerned about how these hours are recorded because an inequity in the pricing and costing of government contracts could occur if uncompensated overtime is worked, but not accounted for, and more than one contract or project is worked on by the salaried employee. The lack of proper accounting for the overtime hours can create the potential for contractors to manipulate their labor accounting systems.

While contractors can choose from a number of different approaches for how to account for uncompensated overtime in pricing proposals, there is the concern that contractors might lower the effective labor rates by proposing uncompensated overtime. In order to satisfy the DCAA and ensure that your proposal demonstrates cost realism, the accounting practices that you use to estimate uncompensated overtime must be consistent with your accounting practices that are used to accumulate and report uncompensated overtime hours. Companies should also include a copy of their uncompensated overtime policy in their proposal.

18.18.2 Addressing Risk in the Cost Volume

While many RFP evaluation criteria place more emphasis on technical and past performance factors over price and the stated evaluation criteria may be "best value," more often than not, cost is a major deciding factor in federal government contract awards. In other words, cost is incredibly important to win. The importance of cost on selection places tremendous importance on pricing strategies and the resultant risks associated with pricing the effort. Some of the critical risks to consider in pricing and the cost volume are shown in Table 18.2.

Table 18.2
Being Aware of Potential Risks Associated with the Cost Volume Helps to Avoid Them

Risk in the Cost Volume	Risk Considerations
Reasonableness	Is our price too high or too low? Will our price be subject to a cost reasonableness analysis performed by the government?
Performance risk	Can we perform the contract for the proposed price?
Cost-accounting strategy risks	A cost-estimating methodology using statistical relationships between historical costs and other program variables such as system physical or performance characteristics, contractor output measures, or staff power loading.
Contract type risks	Cost type contracts shift cost growth risk to the government, while fixed price contracts shift cost growth risk to the contract.
Other risks	Ceilings on indirect rates can impact pricing risk decisions.

18.18.3 What Is Cost Realism?

Cost realism takes into consideration reviewing and evaluating specific elements of each offeror's proposed cost estimate to determine whether the cost elements are realistic for the work to be performed, reflect a clear understanding of the requirements, and are consistent with the unique methods of performance and materials described in the offeror's technical proposal.

Cost realism is not considered in the evaluation of proposals for the award of a fixed-price contract, because these contracts place the risk of loss upon the contractor. However, when awarding a fixed-price contract, an agency may provide for a price realism analysis for the purpose of measuring a company's understanding of the solicitation requirements, or of assessing the risk inherent in an offeror's proposal.

18.18.4 What Is Cost Reasonableness?

According to the FAR, a contract cost is reasonable if, in its nature and amount, it does not exceed that which would be incurred by a prudent person in the conduct of competitive business. In determining the reasonableness of a specific cost, the government will consider:

- Whether it is the type of cost generally recognized as ordinary and necessary for the conduct of the contractor's business or the contract performance;
- The restraints or requirements imposed by such factors as generally accepted sound business practices, arm's-length bargaining, federal and state laws and regulations, and contract terms and specifications;
- The action that a prudent businessperson, considering responsibilities to the owners of the business, employees, customers, the government, and the public at large, would take under the circumstances;
- Any significant deviations from the established practices of the contractor that may unjustifiably increase the contract costs.

18.18.5 What Is Cost or Pricing Data?

In accordance with FAR 15.406-2, federal agencies require cost or pricing data for contracts or subcontract actions where the negotiated values are likely to exceed specific dollar thresholds or meet other specific criteria. When cost or pricing data is required, the contracting officer will request that the offeror complete, execute, and submit to the contracting officer a certification that cost and pricing data are accurate, complete, and current as of a specified date.

Cost or pricing data is defined as all facts that, as of the date of price submission, or, if applicable, an earlier date agreed upon between the parties that is as close as practicable to the date of agreement on price, prudent buyers and sellers would reasonably expect to affect price negotiations significantly. Cost or pricing data consist of more than historical accounting data—facts that can be reasonably expected to contribute to the soundness of estimates of future costs and validity of incurred costs. It includes such factors as vendor quotations, nonrecurring costs, information on changes in production methods and in production or purchasing volume, data supporting projections of business prospects and objectives and related operations costs, unit-cost trends such as those associated with labor efficiency, make-or-buy decisions, estimated resources to attain business goals, and information on management decisions that could have a significant bearing on costs.

18.18.6 Price to Win

Price to win (PTW) has varying definitions in the marketplace. Our definition is that PTW is a function separate from pricing to objectively and independently attempt to determine the likely price, coupled with an anticipated technical score, to win the contract. It can be, by its nature, very subjective. To reduce subjectivity, the collection of data is critical. These data should include the customer's funding, should-cost, and rough order of magnitude, as well as the government's historical buying behavior and buying constraints for this specific opportunity. Likewise, PTW will consider the competition's likely win strategy, approach and teaming strategy, historical pricing behavior, and the competition's likely pricing behavior on the specific opportunity.

In place of PTW, many contractors use competitive analysis (CA). It is similar, but more often performed by the pricer and is based on a common technical solution. Deciding to use CA or PTW is based on resource availability, cost, and the relative value of the PTW information.

The process used to determine the PTW and CA is iterative. It is the responsibility of the capture manager and is designed to increase overall win probability. It cannot be overemphasized how critical a target price is as an overall component in bid/no bid decision.

18.18.7 Common Cost Proposal Problems

Considerable time, effort, and scarce resources are committed to submitting a quality proposal in an effort to secure a contract. You need to ensure that the cost proposal offers the best value to the government, is consistent with the other volumes, and does not contain glaring weaknesses. Some of the more common proposal problems/challenges are:

- Not reviewing the entire RFP from a pricing perspective;
- Not establishing a cost/price volume schedule at the outset;
- Not allowing sufficient time to gather challenging data such as rates from subcontractors and other direct cost quotes;
- Proposals with omissions or failure to follow the RFP;
- Lack of detail supporting indirect rates;

- Lack of narrative with supporting rationale;
- Inconsistent content within the proposal;
- Not using the government-provided pricing model;
- Not ensuring the government-provided pricing model allows for your company's accounting calculations;
- Not submitting the proposal on time;
- Not submitting the complete proposal (including subcontractors);
- Failure to propose correct wage determination minimum rates;
- Professional base rates that are too low or too high;
- Failure to identify uncompensated overtime;
- Government not using the latest DCAA-approved rates;
- Failure to use the correct government estimate for other direct costs;
- Failure to report and provide a cost estimate for additional other direct costs, if appropriate;
- Not submitting a total compensation plan when required;
- Not completing the audit file immediately after submission;
- Not holding a lessons-learned meeting or AAR within a week from submission.

18.18.8 Useful Websites (Note That They May Change) and Templates (Always Check for the Latest Template Online)

- FAR and FAR Supplement Site: https://www.acquisition.gov/browse/index/far;
- Federal Travel Regulations: https://www.gsa.gov/policy-regulations/regulations/federal-travel-regulation;
- Joint Travel Regulations: https://www.travel.dod.mil/Policy-Regulations/Joint-Travel-Regulations/;
- GSA Per Diem: https://www.gsa.gov/travel/plan-book/per-diem-rates;
- System for Award Management: https://sam.gov/content/home;
- Bureau of Labor Statistics: https://www.bls.gov/;

- Department of Labor Wage Determinations: https://sam.gov/content/wage-determinations;
- Certificate of Cost or Pricing Data (Figure 18.2);
- SF 1408 (Figure 18.3).

18.19 PRICE PROPOSAL REFERENCE

Lindquist, M., *Secrets of Strategic Pricing for Government Contractors,* Cave Creek, AZ: Granite Leadership International Corporation, 2022.

CERTIFICATE OF CURRENT COST OR PRICING DATA

This is to certify that, to the best of my knowledge and belief, the cost or pricing data (as defined in section 2.101 of the Federal Acquisition Regulation (FAR) and required under FAR subsection 15.403-4) submitted, either actually or by specific identification in writing, to the Contracting Officer or to the Contracting Officer's representative in support of ______* are accurate, complete, and current as of ______**. This certification includes the cost or pricing data supporting any advance agreements and forward pricing rate agreements between the offeror and the Government that are part of the proposal.

Firm ______________________________

Signature ______________________________

Name ______________________________

Title ______________________________

Date of execution*** ______________________

* Identify the proposal, request for price adjustment, or other submission involved, giving the appropriate identifying number (*e.g.*, RFP No.).

** Insert the day, month, and year when price negotiations were concluded and price agreement was reached or, if applicable, an earlier date agreed upon between the parties that is as close as practicable to the date of agreement on price.

***Insert the day, month, and year of signing, which should be as close as practicable to the date when the price negotiations were concluded and the contract price was agreed to.

(End of certificate)

(b) The certificate does not constitute a representation as to the accuracy of the contractor's judgment on the estimate of future costs or projections. It applies to the data upon which the judgment or estimate was based. This distinction between fact and judgment should be clearly understood. If the contractor had information reasonably available at the time of agreement showing that the negotiated price was not based on accurate, complete, and current data, the contractor's responsibility is not limited by any lack of personal knowledge of the information on the part of its negotiators.

(c) The contracting officer and contractor are encouraged to reach a prior agreement on criteria for establishing closing or cutoff dates when appropriate in order to minimize delays associated with proposal updates. Closing or cutoff dates should be included as part of the data submitted with the proposal and, before agreement on price, data should be updated by the contractor to the latest closing or cutoff dates for which the data are available. Use of cutoff dates coinciding with reports is acceptable, as certain data may not be reasonably available before normal periodic closing dates (*e.g.*, actual indirect costs). Data within the contractor's or a subcontractor's organization on matters significant to contractor management and to the Government will be treated as reasonably available. What is significant depends upon the circumstances of each acquisition.

(d) Possession of a Certificate of Current Cost or Pricing Data is not a substitute for examining and analyzing the contractor's proposal.

(e) If certified cost or pricing data are requested by the Government and submitted by an offeror, but an exception is later found to apply, the data shall not be considered certified cost or pricing data and shall not be certified in accordance with this subsection.

Figure 18.2 Certificate of Current Cost or Pricing Data. (*From:* https://www.acquisition.gov/far/15.406-2. Public domain.)

PREAWARD SURVEY OF PROSPECTIVE CONTRACTOR (ACCOUNTING SYSTEM)	SERIAL NUMBER *(For surveying activity use)*	**OMB Control Number: 9000-0011** **Expiration Date: 1/31/2024**
	PROSPECTIVE CONTRACTOR	

Paperwork Reduction Act Statement - This information collection meets the requirements of 44 U.S.C. § 3507, as amended by section 2 of the Paperwork Reduction Act of 1995. You do not need to answer these questions unless we display a valid Office of Management and Budget (OMB) control number. The OMB control number for this collection is 9000-0011. We estimate that it will take 24 hours to read the instructions, gather the facts, and answer the questions. Send only comments relating to our time estimate, including suggestions for reducing this burden, or any other aspects of this collection of information to: U.S. General Services Administration, Regulatory Secretariat Division (M1V1CB), 1800 F Street, NW, Washington, DC 20405.

SECTION I - RECOMMENDATION

1. PROSPECTIVE CONTRACTOR'S ACCOUNTING SYSTEM IS ACCEPTABLE FOR AWARD OF PROSPECTIVE CONTRACT

☐ YES ☐ NO *(Explain in 2. NARRATIVE)*

☐ YES, WITH A RECOMMENDATION THAT A FOLLOW ON ACCOUNTING SYSTEM REVIEW BE PERFORMED AFTER CONTRACT AWARD *(Explain in 2. NARRATIVE)*

2. NARRATIVE *(Clarification of deficiencies and other pertinent comments. If additional space is required, continue on plain sheets of paper.)*

IF CONTINUATION SHEETS ATTACHED - MARK HERE ☐

3. SURVEY MADE BY	a. SIGNATURE AND OFFICE *(Include typed or printed name)*	b. TELEPHONE NUMBER *(include area code)*	c. DATE SIGNED
4. SURVEY REVIEWING OFFICIAL	a. SIGNATURE AND OFFICE *(Include typed or printed name)*	b. TELEPHONE NUMBER *(include area code)*	c. DATE REVIEWED

AUTHORIZED FOR LOCAL REPRODUCTION
Previous edition is NOT usable

STANDARD FORM 1408 (REV. 1/2014)
Prescribed by GSA - FAR (48 CFR) 53.209-1(f)

Figure 18.3 Standard Form 1408: Preaward Survey of Prospective Contractor Accounting System. (*From:* https://www.gsa.gov/system/files/SF1408-14e.pdf. Public domain.)

SECTION II - EVALUATION CHECKLIST			
MARK "X" IN THE APPROPRIATE COLUMN *(Explain any deficiencies in SECTION I NARRATIVE)*	YES	NO	NOT APPLI-CABLE
1. EXCEPT AS STATED IN SECTION I NARRATIVE, IS THE ACCOUNTING SYSTEM IN ACCORD WITH GENERALLY ACCEPTED ACCOUNTING PRINCIPLES APPLICABLE IN THE CIRCUMSTANCES?			
2. ACCOUNTING SYSTEM PROVIDES FOR:			
a. Proper segregation of direct costs from indirect costs.			
b. Identification and accumulation of direct costs by contract.			
c. A logical and consistent method for the allocation of indirect costs to intermediate and final cost objectives. (A contract is final cost objective.)			
d. Accumulation of costs under general ledger control.			
e. A timekeeping system that identifies employees' labor by intermediate or final cost objectives.			
f. A labor distribution system that charges direct and indirect labor to the appropriate cost objectives.			
g. Interim (at least monthly) determination of costs charged to a contract through routine posting of books of account.			
h. Exclusion from costs charged to government contracts of amounts which are not allowable in terms of FAR 31, Contract Cost Principles and Procedures, or other contract provisions.			
i. Identification of costs by contract line item and by units (as if each unit or line item were a separate contract) if required by the proposed contract.			
j. Segregation of preproduction costs from production costs.			
3. ACCOUNTING SYSTEM PROVIDES FINANCIAL INFORMATION:			
a. Required by contract clauses concerning limitation of cost (FAR 52.232-20 and 21) or limitation on payments (FAR 52.216-16).			
b. Required to support requests for progress payments.			
4. IS THE ACCOUNTING SYSTEM DESIGNED, AND ARE THE RECORDS MAINTAINED IN SUCH A MANNER THAT ADEQUATE, RELIABLE DATA ARE DEVELOPED FOR USE IN PRICING FOLLOW-ON ACQUISITIONS?			
5. IS THE ACCOUNTING SYSTEM CURRENTLY IN FULL OPERATION? (If not, describe in Section I Narrative which portions are (1) in operation, (2) set up, but not yet in operation, (3) anticipated, or (4) nonexistent.)			

STANDARD FORM 1408 (REV. 1/2014) **BACK**

Figure 18.3 *(continued)*

19

ACADEMIC AND GOVERNMENT GRANT PROPOSALS AND INTERNATIONAL AND PRIVATE-SECTOR PROPOSALS

19.1 FEDERAL GRANTS

A federal grant is a way that the U.S. government funds your ideas and projects to provide public services and stimulate the economy. Grants support critical recovery initiatives, innovative research, and many other programs listed in the Catalog of Federal Domestic Assistance (CFDA). The grant process follows a linear life cycle that includes creating the funding opportunity, applying, making award decisions, and successfully implementing the award. See GRANTS.GOV for lots of additional information and guidance. Federal funding opportunities published on GRANTS.GOV are for organizations and entities supporting the development and management of government-funded programs and projects.

19.2 FEDERAL CONTRACTS

A federal contract is a legally binding agreement in which federal agencies request certain terms and conditions in order to acquire a desired service or good furnished by the selected awardee. The awardee is selected by demonstrating the best proposal to achieve the service or good, both technically and economically. A federal contract is used where the principal purpose is to acquire property or services for the direct benefit of the U.S. government (31 USC 6303). In contrast, the U.S. government uses grants to carry out a public purpose of support or stimulation (31 USC 6304). Accordingly, grants are more flexible in their terms and conditions than contracts since the intent with grants is more general in nature.

Contracts are governed by relatively strict terms and conditions (Ts & Cs), including clauses from the FAR. Grants are governed by the terms of the Notice of Award (NOA).[1]

The Code of Federal Regulations (CFR) Title 2 contains the codified federal laws and regulations that are in effect as of the date of the publication pertaining to Federal Grants and Agreements. The legislative governance for federal contracts is the FAR. For federal grants, the Office of Federal Financial Management (OFFM) provides Office of Management and Budget (OMB) oversight, whereas for FAR contracts, the Office of Federal Procurement Policy (OFPP) provides that oversight.[2]

19.3 COMPARISON OF GRANT PROPOSAL WRITING AND COMPETITIVE FEDERAL PROPOSAL WRITING

19.3.1 Grant Proposal Writing

- Select a compelling and appropriate topic of proper scope (not too all-encompassing; in effect, do not try to "boil the ocean").
- Ensure that the research or activity proposed will advance the grantmaking organization's mission and objectives (e.g., Bank of America Charitable Foundation Inc. or The Wells Fargo Foundation).
- Adhere to the structure of the grant application.
- Demonstrate the ability to measure and report on established outcomes-based success metrics aligned to strategic objectives.

1. https://your.yale.edu/research-support/office-sponsored-projects/contracts/federal-contracts (Yale University).
2. https://research.wustl.edu/identify-a-federal-contract/ (Washington University in St. Louis).

- Demonstrate the scientific or technical merit, and programmatic merit, of the research or activity. Validate with quantitative details.
- Place the research or activity into a larger societal context and demonstrate how it will contribute to, for example, lives being improved, jobs being created, or the environment being preserved.
- Demonstrate how the project fits within the published academic and professional literature and what gap it will address.
- Describe the research methodology and identify key assumptions.
- Provide a detailed work plan that describes who does what, how, when, and with whom, and for how much money. Provide a detailed budget with proper justifications.
- Include a plan for instruction in ethics and responsible conduct of the research.
- Provide a data management plan.
- Be certain to provide hard-hitting descriptions of the relevant experience, academic credentials, and publications of the principal investigator (PI), co-PI, and other key personnel, such as the data scientist.

19.3.2 Competitive Federal Proposal Writing

- Adhere to the structure of Section L, Section M, and Section C of the RFP or RFQ.
- Convey detailed understanding of the work to be performed and the environment in which it will be accomplished. Describe the work to be performed within a larger business and mission context.
- Articulate a clear approach to meeting the requirements conveyed in the SOW and PWS, as well as in the Quality Assurance Surveillance Plan (QASP) or Performance Requirements Summary (PRS).
- Provide validation for the approach by reference to proof points gathered from past performance.

- Highlight tangible and intangible strengths that the proposing company's approach will offer the government in terms that the agency cares about (focus on Section M; leverage business intelligence gained through customer interactions). Validate with quantitative details.
- Provide results-focused résumés that focus on how the particular key personnel staff professional is prepared to meet technical, programmatic, and operational cadence changes that will occur during the period of performance.

19.4 INTERNATIONAL COMMERCIAL PROPOSAL DEVELOPMENT

International proposals of all stripes require the offeror to provide an approach and a clear indication of the value that the approach provides. It is important that the international business that issued the solicitation document can determine the extent to which the stated requirements have been met.

Differences mean professional possibilities. Cultural differences, native language differences, and time zone differences are all parts of international commercial proposal development. Working with culturally diverse technical SMEs across the world presents meaningful opportunities (and challenges) for proposal professionals in the United States. How? They must help translate domain-specific, quantitatively focused solutions into a hard-hitting business case.

Unlike U.S. federal government competitive procurements, commercial business-to-business (B2B) proposals for overseas clients can involve several F2F interactions with the client's management and technical teams subsequent to the release of the final solicitation documents. This presents multiple opportunities to successively enhance and tune the management and technical solution sets. Given the 24/7 work cycle and client expectations for rapid modifications and iterations, the velocity of this process makes the international commercial proposal development environment particularly invigorating. It also becomes quite demanding from perspectives such as communication and document configuration control.

To be sure, there are considerable similarities between international commercial B2B solicitations and U.S. federal competitive

RFPs. In a recent $100 million commercial request for solution, for instance, there were numbered sections that addressed format, page count, and file-naming conventions—similar to Section L of the Uniform Contract Format (UCF) of the FAR. In addition, there were evaluation criteria and considerations nearly identical to Section M.

Further, there was also discussion of quality, capability, cost management, approach, and value to be delivered—clearly recognizable language from the U.S. federal marketspace. Of note in the commercial RFS was special focus on "ensuring that the relationship and operational interaction with [the client] will be well managed." Clearly, the long-term interaction between offeror and ultimate client was envisioned by that client to extend far beyond a transactional level.

Data analytics, Agile data stores, Lean principles, Centers of Excellence, and robotics were important elements in the technical solution set for this particular commercial proposal. However, the pivotal decision makers of the ultimate client—a multibillion-dollar, European-based corporation with a global presence—directed laser-like attention toward the business side of the equation. They zeroed in on the business value to be delivered by the talented people, advanced processes, deep knowledge, and leading-edge tools presented in the offeror's proposal. Indeed, there was an entire series of pre- and post-RFS-release meetings that focused almost exclusively on the business case and the sustainability of the potential long-term partnership.

Therefore, technical details—which certainly needed to be presented and illustrated—had to be transformed into business benefits. These encompassed compressed time-to-value horizons, decreased total cost of ownership (TCO), increased corporate-level visibility into the program, relevant value levers and when to activate them, and lower risk profiles, as well as support for triple bottom line (TBL) business practices. TBL addresses the social dimensions of a corporation. Also, the proposal had to demonstrate how the ultimate client would move up the value chain within its particular industry, based upon the solutions being offered. Conceptualized by Dr. Michael E. Porter of the Harvard Business School, value chain refers to changing business inputs into business outputs such that they have greater value than the original cost of creating those outputs.

One mechanism that can be used to show that a team has carefully considered multiple hypotheses and what-if scenarios is a table with decision variables, options, and implications as the three column headings. This can be a powerful tool for demonstrating understanding of the ultimate client's business environment, both today and in the future. This technique can be applied in U.S. government proposals as well.

Among the valuable lessons learned:

- Appoint two document owners for each proposal module—one in the United States and the other in India or South Africa, for example, to cover the entire 24-hour cycle in a given day. These document owners will be fully accountable for configuration (version) control.
- Introduce all core team members early in the proposal life cycle, to facilitate full understanding of roles and decision-making authority, "swim lanes," plans of action, and milestones.
- Conduct regular stand-up videoconferences or teleconferences to communicate progress, priorities, critical near-term issues, and the geographic location of core team members so everyone knows in which time zones their colleagues are working. Overcommunication is critical to success within distributed teams.
- Develop the end-to-end proposal outline early in the process, and continually validate it against the solicitation document and the client's verbal directions, as they evolve over time, to ensure exacting and full compliance.

20

EPILOGUE

20.1 DIRECT BENEFITS OF FEDERAL SUBCONTRACTING GOALS, STRATEGIC PARTNERING, AND MENTOR-PROTÉGÉ RELATIONSHIPS

In 2023, the opportunities for small business success have never been better. Currently, the federal government's subcontracting goals[1] stand at:

- 23% for small businesses;
- 5% for small disadvantaged businesses (SDBs);
- 5% for women-owned businesses (WOBs);
- 3% for HUBZone (Historically Underutilized Business Zones) small businesses;
- 3% for service-disabled veteran-owned businesses.

These guidelines are of direct benefit to small businesses of all types because they require large federal contractors to proactively team with small businesses to pursue U.S. government contracts. Small business entrepreneurs can significantly leverage their limited

1. Congressional Research Service, "Federal Small Business Contracting Goals," September 25, 2022, https://crsreports.congress.gov/product/pdf/IN/IN12018.

time, marketing, financial, and technical resources, as well as augment their revenue stream. How? By seeking and forming strategic partnerships as subcontractors and as protégés with large prime contracting companies to pursue new business opportunities and expand work with existing customers.

As a small business, you will want to learn how to market your firm to large business primes in ways that extend far beyond meeting a socioeconomic category in a small business utilization plan. When you have the opportunity to respond to a prime's administrative data calls, ensure that you are prompt and provide complete information so that the likelihood of your inclusion on the prime's team goes up.

In addition, understand that your prime contractor partner will have to sell your small company's human talent and knowledge base, contractual experience, and fiscal solvency in its proposal to a given government customer. Make that task extremely easy for the prime by having a customer-focused, fact-driven story to tell, and providing that information in a timely manner during the proposal response life cycle. Far too many small businesses submit amateurishly written, incomplete materials to the prime contractor for integration into the prime's proposal. This ineffective practice adds to the prime contractor's bid and proposal (B&P) costs and detracts from your working relationship with that prime in the future. Learn how to conduct contractual business with the prime and the ultimate government customer. Demonstrate your company's ability to staff the project with stellar professionals who are committed to the life of the project and not merely senior staff whose résumés help to win the job, but who then leave the project after a minimal amount of time. Build trust and camaraderie in the relationship and establish a strategic partnership for the long term.

Critical to a small firm's subcontracting and partnering success with a large prime contractor are the following elements: (1) your superlative contractual performance on existing or past projects; (2) your flexible, audited, and approved pricing strategies and structures; (3) the allocation of your top-flight professional staff to participate in proposal development and on review (and possibly oral presentation) teams with the prime contractor; (4) your fiscal strength as measured by such parameters as operating profit, net income, positive cash flow, operating history, and future business assumptions; (5) your well-defined and articulated core competencies and product/

service lines; and (6) your fair, equitable, and ethical teaming agreements that are honored, always.

20.2 SPECIFIC STRATEGIES FOR ACHIEVING FEDERAL AND PRIVATE-SECTOR SUBCONTRACTS

- Register your small business with sam.gov (https://sam.gov/content/home).
- Mentor-protégé programs can help your business. The DoD Mentor-Protégé Program (MPP) was enacted in 1990 (Public Law 101-510) under the direction of former Senator Sam Nunn and Secretary of Defense William Perry. In the past 5 years, the DoD's MPP has successfully helped more than 190 small businesses fill unique niches and become part of the military's supply chain.[2] Other federal agencies also have Mentor-Protégé programs, including the Small Business Administration (SBA), NASA, and the Department of Homeland Security (DHS), as well as the Department of Transportation (DOT) and the Department of Energy (DOE). In addition, the U.S. Department of Agriculture (USDA) maintains a National Mentoring Program (NMP).
- Attend and participate in appropriate government and industry conferences and tradeshows. For example, the SBA conducted its 6th Annual Mentor Protégé Conference in March 2023. The DOE convened its Small Business Forum and Expo in July 2023. The DoD presented its Mentor Protégé Summit 2023 in March 2023. Federal Publications Seminars (FPS), a leader in government contracts training, hosted its 2023 Small & Medium Business GovCon Forum in July 2023.
- Register and prequalify your small business with the SADBUS, or Small and Disadvantaged Business Utilization Specialist, in specific federal agencies with which your company wants to do business. At the state level, the analogous office may be called the Office of Supplier Diversity and Outreach. Many large corporations now have small business centers, staffed

2. https://business.defense.gov/Programs/Mentor-Protege-Program/.

with Small Business Liaison Officers (SBLOs). Every federal contracting agency and prime contractor doing substantial business with the federal government is obliged to designate an SBLO or SADBUS within the agency or firm. This person is the first point of contact for your small business wishing to make your products and services known to the federal government contracting entity.

- Interact with Small Business Development Centers (SBDCs) in your state. The 62 SBDCs with 1,000 service locations nationwide are educational and research resources for small businesses. In addition, some states maintain Regional Minority Supplier Development Councils and Regional Minority Purchasing Councils.
- Register your company in the Diversity Information Resources (DIR) Online Supplier Sourcing Portal (http://www.diversityinforesources.com/). It is free and makes your company's qualifications readily available to DIR's network of corporate buyers.
- Proactively participate in Private-Sector Supplier Diversity Programs. According to JP Morgan,[3] each member of the Billion Dollar Roundtable (BDR)—composed of 28 *Fortune* 500 companies—spent at least $1 billion annually with minority-owned and woman-owned businesses. In May 2023, WEConnect International convened its 2023 Global Symposium for Supplier Diversity and Inclusion in Washington, D.C. BMW sponsored its 2023 BMW Supplier Diversity Conference in August 2023 in South Carolina. The 2023 National Small Business Conference in New Orleans in February 2023 was a supplier diversity networking event for small businesses. The National Minority Supplier Development Council (NMSDC) held the NMSDC Annual Conference & Exchange 2023 in October 2023 in Baltimore, Maryland.

3. "Diverse Suppliers Are Becoming Crucial Links in Corporate Supply Chains," https://www.jpmorgan.com/commercial-banking/insights/big-companies-back-supplier-diversity#:~:text=Diverse%2DSupplier%20Pie%20Keeps%20Growing&text=Leaders%20in%20this%20trend%20include,%2D%20and%20woman%2Downed%20suppliers.

Strategic alliances and partnerships must be managed as enduring business relationships—with mutually compatible objectives, shared risks and resources, real-time knowledge transfer, and trust. Success will follow.

20.3 BEST-PRACTICE B&P SCENARIO—SHIFT LEFT

To help optimize the probability of winning, we need to reexamine the trajectory and cadence of B&P investment for a specific federal pursuit. Particularly within small federal support services contractors, B&P dollars are quite limited, and so are business development, capture management, and proposal development resources and technical SMEs. The red solid line in Figure 20.1 represents the typical B&P spending curve within small businesses. Note that during the pre-final RFP release phase of the proposal development life cycle, the level of effort (LOE) expended on the specific government pro-

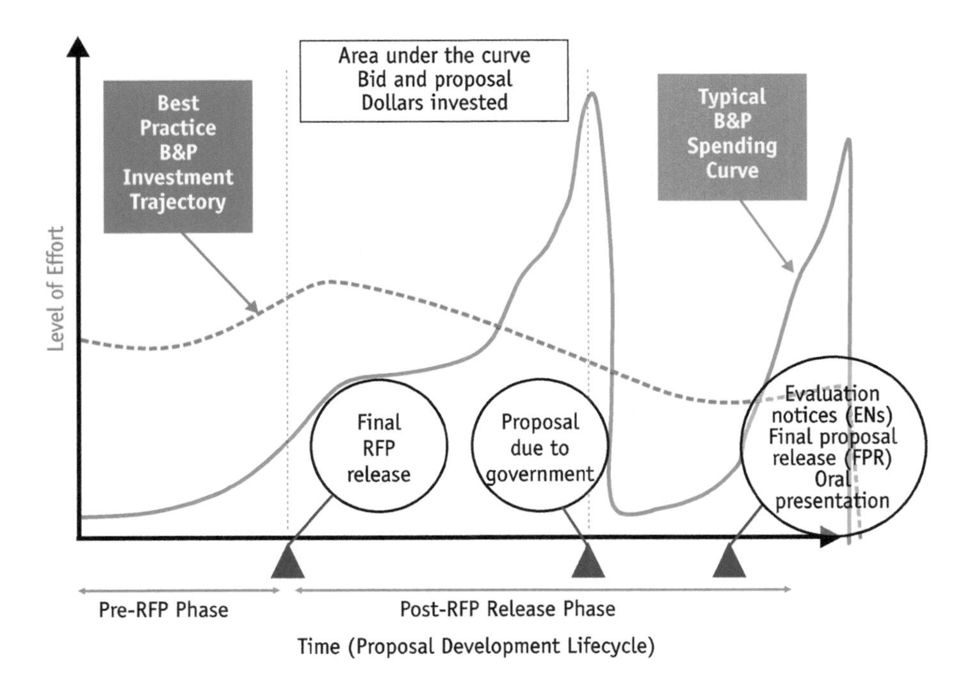

Figure 20.1 Best practice B&P scenario. (Conceptualized by Dr. R. S. Frey.)

gram pursuit is relatively low. This translates directly to few B&P dollars being invested, along with suboptimal preparation.

Conversely, the green dashed line associated with the best practice B&P investment trajectory depicts a left-shifted LOE and investment strategy. Ultimate proposal success is the direct result of informed bid/no-bid decisions and lots of upfront work, including pursuit strategy planning and development, interaction with the government decision-makers and influencers, intelligent teaming, strengths development and validation, and advanced solution development, to name a few.

When the government releases the final RFP, the slope of the red solid line is positive and quite steep. It crescendos during the final weeks of the proposal development life cycle. In many instances, a small company's B&P budget gets tossed aside in order to throw resources at the situation. Extra consultants and freelance proposal professionals are engaged at premium rates. Often there is significant rework. Then, once the proposal is submitted to the government, B&P expenditures fall off precipitously.

The slope of the green dashed line is actually somewhat negative during the proposal response period. Advanced work and preplanning have reduced the need for frantic spending. To be sure, there are still B&P dollars being invested, but at a slightly declining rate. In preparation for responses to evaluation notices, FPR, and oral presentations, the B&P investment level remains strong, but does not exhibit sharp spikes. This makes financial planning much more predictable and fiscal control more effective.

The red solid line goes up sharply once again when ENs, FPRs, and oral presentations occur. Then the level of B&P investment often goes to zero after final submittal of all proposal deliverables. However, the green dashed line continues in order to accommodate an AAR, or retrospective, in Agile parlance.

The area under the red curve, that is, the total B&P dollars invested, is most often much higher than under the green curve. Applying the best practice B&P investment trajectory contributes to high Capture and Proposal Team morale and strong levels of staff retention, reduced proposal rework, and increased proposal throughput per unit time.

20.4 SO, YOU ARE JUST GETTING STARTED WITH PROPOSALS

I have been asked about what guidance I would offer to an individual just starting out in proposal development. First and foremost, become an information sponge (see Figure 20.2).

Learn about the federal acquisition life cycle or commercial client's buying cycle, as well as your organization's business development, capture management, and proposal development staff, processes, and terms of art. For proposal work within the U.S. federal government marketspace, become very familiar with the 13 parts (Sections A to M) that comprise a federal RFP, which follows the FAR Uniform Contract Format (UCF). Most federal agencies do adhere to the UCF. Importantly, join and participate in the structured activities of your local chapter of the Association of Proposal Management Professionals (APMP).[4] Work to build social relationships with fellow proposal professionals.

Figure 20.2 Known as the "Sponge Capital of the World," Tarpon Springs, Florida, is an area where Greek immigrants settled during the early 1900s. (Photograph © Dr. R. S. Frey.)

4. https://www.apmp.org/.

Invest time learning about exactly what your company is selling. For example, GE focuses on "building a world that works." CACI transforms "actionable data into information advantage." SAIC accelerates "outcomes." Universities Space Research Association (USRA) advances "space science and technology." Become conversant in your firm's key technical support services offerings, and know who your primary federal or commercial customers are.

In addition, become intimately familiar with the processes and systems (i.e., technology and software) adopted for use by your company. Generally, small business firms lack market differentiation and SMEs and grapple with high overhead labor costs. Systems and processes can help with market differentiation. Small business firms can also leverage software tools that deliver automation and workflows as a force multiplier. When combined, these systems and processes can assist with controlling (i.e., limiting) overhead labor costs in ways that lead to competitive pricing and winning more contracts.

Recognize that proposal development is absolutely not a 9-to-5 position, and that is okay. Your focus should be on compressing the time to value for yourself as a member of the proposal team. Life does become driven by acquisition cycles. Frequently, federal government procurements shift to the right (in effect, are delayed). It will be advantageous to you to become extremely time-flexible. However, it is vital for you to engage in some type of outside-of-work activity just for you. My personal passions are photography and woodworking, particularly woodturning on my mini-lathe (Figure 20.3). They provide a counterbalance to the demands of proposaling in today's hypercompetitive marketspace. Over the years, I have heard many proposal professionals say that they have no hobby or that hobbies are for retirement. Take care of yourself along the way. Hobbies will contribute to your mental and emotional well-being.

Oftentimes in small organizations, proposal development professionals are faced with a challenge. That hurdle involves generating responses to federal or private-sector solicitation documents without having detailed insights from business development or capture management staff regarding the customer's mission, operational environment, relevant technologies, or governance framework. Do not even expect to have these insights hand-delivered to you.

Becoming a proposal journalist will help you to get smart on your own, surmount these obstacles, and contribute to an increased

Figure 20.3 Lidded cup that I turned on my mini-lathe. It began as six short boards of exotic hardwoods. (Photograph © Dr. R. S. Frey.)

probability of winning (P_{win}). There are a plethora of Web-based resources available to build your customer-specific and technology-specific competencies and confidence. This is not to say that you can Google your way to sustained proposal success. Nonetheless, through mining relevant mission statements, core values, guiding principles and goals, strategic plans, technology roadmaps, agency-specific white papers, handbooks, and Congressional testimony, you can gain an amazing amount of insight, as well as ideas for graphics concepts.

Web-based research on recent awards by a given customer to competitors can also shed considerable light onto that customer's buying habits and preferred technical solutions. Speeches and biosketches or curriculum vitae of key federal or commercial-sector leaders and published papers written by those individuals can prove to be invaluable in proposing meaningful, on-target solutions to those same customer leaders. Given the level of executive review and approval, customer news releases are excellent sources of knowledge. Specific words and phraseology that are used can be incorporated into the proposal or oral presentation.

From a technical and business perspective, there is much to be gained by examining white papers, technical reports, and case studies available from TechRepublic; Ernst & Young; Gartner; KPMG; Booz, Allen, Hamilton; and Advanced Technology Academic Research Center (ATARC) as well as Mitre; Microsoft Azure (white papers, analyst reports, and e-books); AWS (white papers and guides); Mosaic Data Science; and the SANS Institute (cybersecurity white papers).

In addition, websites can help proposal development staff professionals understand how the customer's organization is structured and provide insight into key graphical or word concepts that the customer uses frequently. For example, the Small Spacecraft Systems Virtual Institute (S3VI) publishes a quarterly digest of resources and activities occurring within NASA and external small spacecraft communities that can shed light on the latest points of interest to the customer community. The Federal Emergency Management Agency (FEMA), U.S. Department of Labor (DOL), Occupational Safety and Health Administration (OSHA), and U.S. Environmental Protection Agency (EPA) also have publicly available newsletters.

Documented trip reports based on execution of call plans can be rich reservoirs of direct insight into a customer's hopes, fears, biases, critical issues, and success factors. These trip reports are prepared by members of your company or team who visit with the government customer. Typically, these are stored in a secure knowledge portal such as SharePoint or Google Drive.

Honing your skills as an information miner will yield significant proposal-applicable dividends. Applying a Think→Draw→Write methodology to what you learn will help you to become much more self-sufficient as a proposal professional. Let's take this methodology apart. Think: conduct the research and integrate the findings. Draw: actually draw out key graphics on a whiteboard or on screen with such tools as Microsoft Visio, Gliffy, Google Drawings, and Edraw. Write: develop narrative that addresses the RFP and incorporates elements of what you learned from your research and the call reports.

Current and future proposal management professionals must prepare themselves to be an integral part of the proposal solution set development process. By doing so, they move up the value chain within their organization and become a more sought-after knowledge worker in the marketplace at large. Harvard Business School professor Michael E. Porter described the concept of value chain in his path-breaking book, *Competitive Advantage: Creating and Sustaining Superior Performance.* The value chain in Porter's construct divides a firm into the discrete activities that it performs in designing, producing, marketing, and distributing its product (or services). You will want to expand your skillset and knowledge base to migrate beyond serving in a commodity-type proposal development role that will most likely become automated in the years ahead. There is already RFP and proposal software, such as Qvidian, Relevant Match from Relevant Software Corporation, PerfectIt, and XaitProposal. Look for ways to add value to the services you provide. Earning your PMP certification is one potential avenue. Maturing your craft to a level where you are invited to speak at an APMP Bid & Proposal Conference (BPC) is another pathway to additional value.

I wish you the very best moving ahead with your professional proposal calling. I term it a "calling" because proposaling can be far more than just a job or even a career. It still remains exciting to me after 36 years traveling through proposal-land.

LIST OF ACRONYMS

AAR	After action review
ACM	Association for Computing Machinery
ADDIE	Analysis, design, development, implementation, and evaluation
AFH	*Air Force Handbook*
AI	Artificial intelligence
AICPA	American Institute of Certified Public Accountants
AITP	Association of Information Technology Professionals
ALIC	Archives Library Information Center
AMC	Army Materiel Command
APB	Accounting Principles Board
APM	Agile Project Management
APMP	Association of Proposal Management Professionals
APQC	American Productivity & Quality Center

ARC	Ames Research Center (NASA)
ASPE	Assistant Secretary for Planning and Evaluation
ASTD	American Society for Training & Development
ATARC	Advanced Technology Academic Research Center
AWS	Amazon Web Services
B2B	Business to business
B&P	Bid and proposal
BAA	Broad agency announcement
BANI	Brittle, anxious, nonlinear, and incomprehensible
BD	Business development
BDR	Billion Dollar Roundtable
BDUF	Big design up front
BGOV	Bloomberg Government
BI	Business intelligence
BLS	Bureau of Labor Statistics
BLUF	Bottom line up front
BOE	Basis of estimate
BPA	Blanket purchase agreement
BPC	Bid and Proposal Con (APMP)
BSC	Balanced scorecard
CA	Competitive analysis
CAGE	Commercial and government entity
CAS	Cost Accounting Standards

CBO	Congressional Budget Office
CCNP	Cisco Certified Network Professional
CCO	Chief collaboration officer
CDC	Centers for Disease Control and Prevention
CDRL	Contract data requirements list
CEO	Chief executive officer
CFDA	Catalog of Federal Domestic Assistance
CFR	Code of Federal Regulations
ChatGPT	Chat Generative Pretraining Transformer
CIA	Central Intelligence Agency
CIO	Chief information officer
CLIN	Contract Line Item Number
CMM	Capability Maturity Model
CMMI®	Capability Maturity Model Integration
CMMI-DEV	Capability Maturity Model Integration for Development
CMS	Centers for Medicare & Medicaid Services
CMU	Carnegie Mellon University
CNSS	Committee on National Security Systems
CO	Contracting officer
CompTIA	Computing Technology Industry Association
CoP	Community of practice
Co-PI	Coprincipal investigator
COR	Contracting officer's representative

COTR	Contracting officer's technical representative
CPARS	Contractor Performance Assessment Reporting System
CPI	Continuous process improvement
CPM	Critical path method
CRD	Contractor responsibility determination
CSLI	Center for the Study of Language and Information (Stanford University)
CSP	Content services platform
CTO	Chief technology officer
DARPA	Defense Advanced Research Projects Agency
DCCA	Defense Contract Audit Agency
DCMA	Defense Contract Management Agency
DEI&A	Diversity, inclusion, equity, and accessibility
DHHS	Department of Health and Human Services
DHS	Department of Homeland Security
DIR	Diversity information resources
DID	Data item description
DL	Direct labor
DLA	Defense Logistics Agency
DLR	Deutsches Zentrum für Luft- und Raumfahrt
DMS	Document Management System
DoD	Department of Defense
DOE	Department of Energy
DOI	Digital object identifier

DOJ	Department of Justice
DOL	Department of Labor
DPI	Dots per inch
DPLH	Direct productive labor hour
DRD	Data requirements description
DRFP	Draft request for proposal
DRL	Data requirements list
DTP	Desktop publisher/publishing
EBITDA	Earnings before interest, taxes, depreciation, and amortization
EN	Evaluation notice
EPA	Environmental Protection Agency
ERI	Economic Research Institute
ESR	Executive Solution Review
F2F	Face-to-face
FAA	Federal Aviation Administration
FAIM	Functional area and improvement manager
FAO	Field Audit Office
FAPIIS	Federal Awardee Performance and Integrity Information System
FAR	Federal Acquisition Regulation
FAS	Federal Acquisition Services
FASB	Financial Accounting Standards Board
FBR	Fully burdened rate

FDIC	Federal Deposit Insurance Corporation
FEMA	Federal Emergency Management Agency
FHIR	Fast Healthcare Interoperability Resources
FICA	Federal Insurance Contributions Act
FISMA	Federal Information Security Management Act
FOIA	Freedom of Information Act
FOV	Field of view
FPR	Final proposal revision
FPS	Federal Publications Seminars
FTE	Full-time equivalent
FTR	Federal Travel Regulations
FTRR	Flight Test Readiness Review
FUTA	Federal Unemployment Tax Act
FY	Fiscal year
G&A	General & administrative
GAAP	Generally accepted accounting practices
GAGAS	Generally Acceptable Government Auditing Standards
GAO	Government Accountability Office
GRC	Glenn Research Center (NASA)
GSA	General Services Administration
GSFC	Goddard Space Flight Center (NASA)
GWAC	Government-Wide Acquisition Contract
HD/LD	High-demand/low-density

HR	Human resources
HUBZone	Historically Underutilized Business Zone
ICR	Intelligent Character Recognition
ID	Instructional design
ID/IQ	Indefinite delivery/indefinite quantity
IEC	International Electrotechnical Commission
IM	Instant message
IMP	Integrated Master Plan
IPT	Integrated Project Team
IR&D	Independent Research and Development (also IRAD)
ISC	Integration Support Contract (Air Force)
ISO	International Organization for Standardization
IT	Information technology
ITES-3S	Army Information Technology Enterprise Solutions–3 Services (ID/IQ contract)
ITIL	Information Technology Infrastructure Library
ITSM	IT service management
ITSP	IT security plan
JTR	Joint Travel Regulations
JV	Joint venture
KM	Knowledge management
KMMM	Knowledge management maturity model
KO	Contracting officer (DoD)
KPI	Key performance indicator

KT	Knowledge transfer
LaRC	Langley Research Center (NASA)
LCAT	Labor category
LOE	Level of effort
MA	Management approach
MAC	Multiple award contract
MCRC	Marine Corps Recruiting Command
MIOMP	Mission Integration and Operations Management Plan
ML	Machine learning
ML	Maturity Level (CMMI)
MLB	Major League Baseball
MLS	Major League Soccer
MPP	Mentor-protégé program
MTDC	Modified total direct cost
MVP	Maximum value product
MVP	Minimal viable product; most valuable professional (Microsoft PowerPoint recognition)
NAICS	North American Industry Classification System
NARA	National Archives and Records Administration
NASA	National Aeronautics and Space Administration
NBA	National Basketball Association
NCR	No carbon required
NFL	National Football League
NHL	National Hockey League

NIST	National Institute of Standards and Technology
NLP	Natural Language Processing
NMSDC	National Minority Supplier Development Council
NMP	National Mentoring Program
NNSA	National Nuclear Security Administration
NOA	Notice of Award
NOFO	Notice of Funding Opportunity (USAID)
NPA	Network Professional Association
NPD	NASA Policy Directive
NWSL	National Women's Soccer League
OCI	Organizational conflict of interest
ODC	Other direct cost
OFFM	Office of Federal Financial Management
OFPP	Office of Federal Procurement Policy
OKM	Organizational Knowledge Management
OKR	Objective and key result
OH	Overhead
OMB	Office of Management and Budget
OSHA	Occupational Safety and Health Administration
OT	Operational technology
OTA	Other Transaction Authority
PA	Process Area (CMMI-DEV)
P.A.Q.S.	Primary objective (P), audience (A), questions (Q), and subject matter (S)

PDA	Personal digital assistant
PDCA	Plan-Do-Check-Act
PI	Principal investigator
PIC	Proposal Innovation Center
PKI	Proposal knowledge integrators
PKT	Proposal knowledge team
PMA	President's Management Agenda
PMBOK®	Project Management Body of Knowledge
PMI	Project Management Institute
PMO	Project Management Office
PMP®	Project Management Professional
POA&M	Plan of action and milestones
PoP	Period of performance
PPQ	Past performance questionnaire
PRS	Performance Requirements Summary
pt	Point
PTO	Paid time off
PTW	Price to win
P_{win}	Probability of win
PWS	Performance work statement
Q&A	Question and answer
QA	Quality assurance
QAM	Quality assurance manager

QASP	Quality Assurance Surveillance Plan
QMS	Quality management system
QPR	Quarterly Program Review
RACI	Responsible, accountable, consulted, and informed
RFP	Request for proposal
RFI	Request for information
RFQ	Request for quotation
RFS	Request for solution
RMF	Risk management framework
RMP	Risk management plan
ROI	Return on investment
R/Q	Responsibility/qualification
RWD	Responsive web design
S3VI	Small Spacecraft Systems Virtual Institute
S&H	Safety and health
S&T	Science and Technology Directorate (Department of Homeland Security)
SADBUS	Small and Disadvantaged Business Utilization Specialist
SAFe®	Scaled Agile Framework
SAM	Successive Approximation Model; System for Award Management
SAR©	Situation, action, results
SB	Small business
SBA	Small Business Administration

SBDC	Small Business Development Center
SBIR	Small Business Innovation Research
SBLO	Small business liaison officer
SDB	Small disadvantaged business
SDVOSB	Service-Disabled Veteran-Owned Small Business
SEB	Source Evaluation Board
SEI	Software Engineering Institute
SEO	Search engine optimization
SELC	Systems engineering life cycle
SERP	Search engine results page
SESDA	Space and Earth Science Data Analysis (NASA)
SF	Standard Form
SH&E	Safety, health, and environmental
SHRM	Society for Human Resource Management
SLA	Service level agreement
SM	Service mark
SME	Subject matter expert
SOFIA	Stratospheric Observatory for Infrared Astronomy
SOO	Statement of objectives
SOW	Statement of work
SoS	Scrum of Scrums
SP	Special publication (NIST)
SSA	Source Selection Authority

SSDD	Source selection decision document
SSO	Source selection official
SSS	Source selection statement
STAR	Scientific and Technical Aerospace Reports
STEM	Science, technology, engineering, and mathematics
STIF	Scientific & Technical Information Facility (NASA)
SUTA	State Unemployment Tax Act
SWOT	Strengths, weaknesses, opportunities, threats
T&M	Time and material
Ts & Cs	Terms and conditions
T2	Today-tomorrow
TA	Technical approach
TBL	Triple bottom line
TCI	Total cost input
TCO	Total cost of ownership
TCP	Total compensation plan
TCV	Total contract value
TIM	Technical interchange meeting
TINA	Truth in Negotiations Act
TIPI	Technologies, innovations, and process improvement
TM	Trademark
TMG	The McKelvey Group
TORFP	Task Order Request for Proposal

TRL	Technology readiness level
TS	Technical solution (CMMI-DEV)
TTP	Time to productivity
TVA	Tennessee Valley Authority
UC	University of California
UCF	Uniform contract format
UEI	Unique Entity Identifier
USACE	United States Army Corps of Engineers
USAID	United States Agency for International Development
USC	United States Code
USDA	United States Department of Agriculture
USPS	United States Postal Service
USRA	Universities Space Research Association
USTDA	United States Trade and Development Agency
VA	Department of Veterans Affairs
VETS 2	VA Veterans Technology Services 2 (ID/IQ contract)
VMO	Value Management Office
VPC	Virtual Proposal Center
VSM	Value stream map
VUCA	Volatility, uncertainty, complexity, and ambiguity
WBS	Work breakdown structure
WHAC	*W:* What is it? Your core concept and being able to help your audience understand the fundamental elements the way that you do. *H:* How does it work? *A:* Are you

sure? The facts and figures that validate your offering. *C:* Can you do it?

WIP Work in Progress (Agile Project Management)

WMG Western Management Group Government Contractors Compensation Survey

WNBA Women's National Basketball Association

WOB Woman-owned business

WOSB Women-owned small business

WSMR White Sands Missile Range

WTW Willis Towers Watson Data Services

XP Extreme programming

ABOUT THE AUTHOR

Since launching his professional consultancy Successful Proposal Strategies, LLC, in 2007, Dr. Robert S. Frey has helped his customers to win more than $8.39 billion in funded federal government contracts. These contracts span support services engagements across multiple defense, civilian, and intelligence community (IC) agencies. He has supported three winning billion-dollar proposals, two for NASA and one with the Department of Energy (DOE). Of special note is that he has been a substantive contributor as a proposal strategist, architect, writer, reviewer, and orals coach on 22 winning NASA proposals across 5 of NASA's centers nationwide. Dr. Bob has taught NASA's Source Selection process to senior members of a $9.7 billion *Fortune* 500 multinational corporation. In March 2023, he was one of five invited presenters at BAE Systems' virtual Strengths Summit. More than 430 BAE staff from North America, the United Kingdom, and Australia were included.

Dr. Bob brings the insights and lessons learned gained through 36 years of significant, trend-setting, proposal-related accomplishments and thought leadership. This is evidenced by his multiple speaking engagements in both the United States and Canada with the Association of Proposal Management Professionals (APMP), in which he is a Fellow, and the Project Management Institute (PMI), through which he holds his Project Management Professional (PMP®) certification.

More than 4,100 entrepreneurs, small-business and large-business proposal staff, and federal government civil servants have participated in Dr. Bob's proposal training seminars nationwide. In addition, he has conducted in-depth analyses of a broad cross-section of government source selection statements and source selection decision documents.

Dr. Frey also brings broad technical awareness in information technology (IT), engineering, scientific support services, and telecommunications. He is conversant in industry best practices such as The Open Group Architecture Framework (TOGAF) Version 9.1, PMBOK 7th Edition, ITIL V3, CMMI® Institute CMMI Maturity Level 3, ISO 9001:2015 standard, Lean Principles, and Six Sigma.

Growing up professionally within small support services contractors in the federal marketspace, Dr. Bob selected "Leader Self-Efficacy and Resource Allocation Decisions: A Study of Small Business Contractors in the Federal Marketspace" as his doctoral dissertation topic at the University of Maryland University College. His dissertation formed the basis for a work entitled *Increasing Your Success as a Small Business Leader,* which was published by Scholar's Press in Germany. He has continued his formal education through both the University of Denver (Principles of Information Security) and Harvard University Extension School (Agile Project Management (APM), Computer Science for Business Professionals, and Decision Making).

This book, *Successful Proposal Strategies On-The-Go!,* will update the 6th edition of his well-received volume, *Successful Proposal Strategies for Small Businesses,* also published by Artech House. The WorldCat database indicates library holdings for *Successful Proposal Strategies for Small Businesses* extend to 66 countries and provinces worldwide.

INDEX

71% effect, 30–31

A

Abstracting, in proposal writing, 171–72
Accounting system adequacy, 223–24
Accuracy, in proposal writing, 168–69
Acronyms
  defining, 179–81
  list, this book, 259–73
Action captions, 175–76
Active voice, 172–75
ADDIE (Analysis, Design, Development, Implementation, and Evaluation), 209
After action reviews (AAR), 24, 210, 235, 252
Agile-based methodologies, 211
Agile Manifesto, 201, 211
Agile Practice Guide, 144
Agile Project Management (APM)
  application through Scrum, 209
  in context, 210–11
  as cutting-edge management approach, 201
  knowledge sharing and, 43–45
  "retrospective," 210
  sustained engagement and, 209–10
  timeboxing and, 209
Analysis, Design, Development, Implementation, and Evaluation (ADDIE), 209
Artificial intelligence (AI), 47–48
Audit files, 230–31

B

Balanced scorecard (BSC), 30
Bases of estimates (BOEs), 150
"Beginner's Mind," 108
Best practices
  B&P, 251–52
  CMMI, 156
  remote collaboration, 123–24
Bid and proposal (B&P)
  best practice scenario, 251–52
  budget, 252
  costs, 168, 248
  investment trajectory, 252
Bid/no-bid decision-making strategy, 61
Big design up front (BDUF), 209
Billion Dollar Graphics (Parkinson), 145–46
Black Hat reviews, 16–17
Blanket Purchase Agreement (BPA), 191
Blue Team Review
  PMBOK framework for, 22–24
  Proposal Readiness Work Products and, 207
Bonus program, 3–6
Box-in-a-box model, 137–39
Brain-writing, 141
Brevity, in proposal writing, 169
Brittle, anxious, nonlinear, and incomprehensible (BANI), 35
Business acquisition as formal process, 78–80
Business development (BD)
  all personnel are in, 16–18
  in a blanket, 14–16

Business development (BD) (*continued*)
function, 151
"funnel," 17
INPUTS, 19
OUTPUTS, 20
processes, 10
TOOLS & TECHNIQUES, 20
Business processes, making them stand out, 85–86

C

Candidates
evaluation factors language, 56–57, 58, 59
instructions to offerors language, 56
Section L language, 57, 58
Capability Maturity Model (CMM), 46
Capability Maturity Model Integration (CMMI)
best practices, 156
certification, 164
checklist, applying, 136–37
for Development (CMMI-DEV), 136
Maturity Level (ML) 5, 189
Capture management
in a blanket, 14–16
function, 151
INPUTS, 20, 21
OUTPUTS, 22
PMBOK in focusing on, 20–22
TOOLS & TECHNIQUES, 21
Capture manager, 97
Catalog of Federal Domestic Assistance (CFDA), 241–42
ChatGPT, 47, 96
Chief collaboration officer (CCO), 11
Clarity, in proposal writing, 169
Code of Federal Regulations (CFR), 242
Collaboration tools, 39–41
Commercial and Government Entity (CAGE) code, 155
Communication
challenges, 34
as critical success factor, 86–88
effective, 33–34
importance of, 76
informal, support for, 88
pathways, 31
of requirements and timelines, 114–15
trust and, 86
Communities of practice (CoP), 24, 40
Competitive analysis (CA), 234
Competitive federal writing, 243–44
Competitive price, 75
Completeness, Section M and, 127
Compliance
importance of, 75
moving beyond, 61–62
proposal development and, 62
straightforward, as challenging, 64–66
Compliance matrix
appearance of, 67
building, 66–69
DoD and, 66–67
elements, 66
NASA and, 67
Comprehensiveness, Section M and, 129–30
Comprehensive risk management approach, 88–89
Content Services Platforms (CSP), 24
Contract Data Requirements List (CDRL), 66
Contract line item numbers (CLINs), 66
Contractor responsibility determinations (CRDs), 90–91
Contractors
addressing critical requirements and, 90
federal support services, 41, 74
government awards to, 14
prime, 8, 248
"responsible," 89–90
Corporate experience, Section M and, 129
Corporate technology officer (CTO), 190
Cost Accounting Standards (CAS), 215, 222–23
Cost and price volume
aligning with technical and management proposals, 226–27
narrative and production, 229–30
risk in, 231–32
Cost data, 233, 237
Cost estimating methods, 227, 228
Cost objective, 217
Cost proposal problems, 234–35
Cost realism, 232
Cost reasonableness, 232–33
Costs
bid and proposal (B&P), 248
direct, 217
G&A, 220–21
indirect, 218–19
other direct (ODCs), 218

overhead (OH), 219–20
procurement, 220
unallowable, 221–22
Win Strategy white paper and, 150
Cover
goal of, 159
illustrated example, 161
importance of, 159–62
Creativity, Section M and, 128
Credentials, 243
Criteria
selection, 137
stated, 51–52
unstated, 51–55
Critical Path Method (CPM), 24
Cross-walk table, 68
Customer Information Control System (CICS), 106
Customers, understanding, 75

D

Data Item Descriptions (DIDs), 66–67
Data management methods, 94
Data warehouses, 24
Defense Contract Audit Agency (DCAA)
about, 215, 224–25
areas of emphasis, 225
main service, 225
preaward accounting survey, 224
Direct costs, 217
Direct labor, 217, 229
Discriminators, 97
Diversity Information Resources (DIR) Online Supplier Sourcing Portal, 250
Document AI, 48
Document Management Systems (DMS), 24
Documents
configuration control, 114
proposals as knowledge-based, 167–68
source-selection, 187
storage and archiving, 96
Dovetailing, 181–83
DRAFT RFP (DRFP), 9

E

EBuy, 27
Effectiveness, Section M and, 127
Efficiency, Section M and, 128
Evaluation notices (ENs), 10, 252
"Evidence of strengths," 199
Executive summary
precursor to, 147–53
as a requirement, 152–53
targeted, 76
"Win Strategy white paper" and, 147–52

F

Face-to-face (F2F) working session, 2, 9
Familiarity, Section M and, 129
Feasibility, Section M and, 128
Federal Acquisition Regulation (FAR)
about, 3, 13
overview, 215–16
purpose of, 216
Federal Acquisition Service (FAS), 157
Federal Awardee Performance and Integrity Information System (FAPIIS), 90
Federal contracts, 241–42
Federal Government Proposal Writing (Brown), 178
Federal grants, 241
Federal Information Security Management Act (FISMA), 138
Federal Publications Seminars (FPS), 249
Federal subcontracting goals, 247–49
Federal Travel Regulations (FTR), 222
Fee, 221, 229
Field of view (FOV), 200
Final cost objective, 217
Final proposal revision (FPR)
B&P and, 252
meetings, 10
submission, 1–2
Fonts, 63–64
Framing
proposal development and, 83–85
resources, 84–85
theory, 83
Freedom of Information Act (FOIA), 21
Fringe benefits, 219
From PMO to VMO (Augustine, Cuellar, and Scheere), 143
Full-time equivalents (FTEs), 9
Fully burdened rate (FBR), 221

G

G&A costs, 220–21
Gap analysis, 9–10
Generally accepted accounting practices (GAAP), 223

General Services Administration (GSA), 157
Go/no-go assessment, 75
"Good is the enemy of great," 73–75
Government Accountability Office (GAO), 126, 163–64
Government awards to contractors, 14
Government-wide acquisition contract (GWAC), 5, 27, 191
Grant proposal writing, 242–43
GRANTS.GOV, 241
Graphics
  about, 10, 76
  bid quality and, 159
  communication, 161–62
  completeness, reasonableness, effectiveness, efficiency and, 128
  decipherable, 91
  drawing out, 256
  example of, 160
  importance of, 157–59
  overseeing, 114

H

Horizontal reviews, 195–97
How to Play the Federal Contractor Game to Win (Kritzer), 143–44
HUBZone small businesses, 14, 247
Human dynamics
  communication and, 33–34
  importance of, 27–36
  meetings and, 32–33
  metaphors and, 27–30
  organizational culture and, 34–36
  71% effect and, 30–31
Hy Silver Proposal Writing Methods, 175

I

Implementation roadmap, 137
Incentive programs, 86
Indefinite delivery/indefinite quality (ID/IQ) opportunities, 5
Independent research and development (IRAD), 190
Indirect costs
  about, 218–19
  calculation process, 228–29
  estimation of, 228–29
  See also Costs
Information collection practice, 135–36
Innovation, Section M and, 128
INPUTS
  in business development, 19
  for capture management, 20, 21
  in knowledge management (KM), 24
  for successful Blue Team Review, 22
Integrated project team (IPT), 150
Integration Support Contract (ISC), 174
International commercial proposal development
  about, 244
  B2B solicitations similarities, 244–45
  lessons learned, 246
  multiple hypotheses, 245–46
  technical solution set, 245

J

The Jelly Effect (Bounds), 141
Job embeddedness, 17–18
Joint ventures (JVs), 2–3

K

Key performance indicators (KPIs), 120
Knowledge codification, 24
Knowledge management (KM)
  about, 40
  Agile Project Management (APM) and, 43
  artificial intelligence (AI) and, 47–48
  environments, 43
  fact-based bid/no-bid and, 16
  focus on, 10
  function, 16–17, 46
  INPUTS, 24
  maturity frameworks, 46
  organizational (OKM), 40
  OUTPUTS, 25–26
  PMBOK framework for, 24–26
  processes, 45
  TOOLS & TECHNIQUES, 24–25
Knowledge management maturity model (KMMM), 46
Knowledge visualization, 24–25

L

Labor categories (LCATs), 218
Labor escalation, computing, 227–28
Leadership, 1–3, 8, 17, 22, 29–30, 32, 42, 46
Lean principles, 24
Level of effort (LOE), 251–52
Line spacing, 63–64
LinkedIn profiles, 164–65

M

Made to Stick (Heath and Heath), 145
Making Knowledge Management Clickable (Hilger and Wahl), 142–43
Management approach (MA)
  about, 150
  aligning cost proposal with, 226–27
  elements, 82
  questions, 53–54
  Win Strategy white paper and, 148
Maximum value product (MVP), 210
Meetings
  action items, tracking, 32–33
  facilitator, 119–20
  mutual respect and, 119
  remote collaboration, 124
  rules of engagement, 119–20
  scheduling, 32
  timeboxing and, 33
Mentor-protégé programs, 247–49
Merit, Section M and, 130
Metaphors, 27–30
Mission quality factor, 192
Modified total direct cost (MTDC), 221
More Solutions for Winning the Federal Contractor Game (Kritzer), 144
Multiple award contract (MAC), 5

N

Notice of Award (NOA), 242

O

Objectives and key results (OKRs), 174
Office of Federal Procurement Policy (OFPP), 13–14
Onboarding process, 86
On-the-Go!, xviii
Oral presentations
  optimal presenters, selecting, 92–93
  resources, 92
  virtual, best practices, 91–92
Organizational Conflict of Interest (OCI) plan, 97
Organizational culture, performance success and, 34–36
Organizational KM (OKM), 40
Other direct costs (ODCs), 218
Outline, proposal
  building, 95
  in proposal directive, 112
  Section L and, 182
OUTPUTS
  in business development, 20
  for capture management, 22
  in knowledge management (KM), 25–26
  for successful Blue Team Review, 23–24
Overhead (OH), 219–20

P

Passion
  about, 1
  focused, 8–9, 27
  importance of, 1, 6
  inspiring, 11
  in oral presentations, 91–92
  vectored, 77
Past performance, 76, 148, 150, 192
Perception support, 201
Performance confidence assessment, 126
Performance Requirements Summary (PRS), 243
Performance work statement (PWS)
  about, 7, 20
  approach to meeting requirements in, 243
  mapping Section L elements to, 66
  technical approach (TA), 128
  unstated criteria and, 52
Period of performance (PoP), 120
Persuasive Business Proposals (Sant), 178
Phase-in/phase-out plan, 148, 150
Plan-Do-Check-Act (PDCA), 211
Plan-Do-Study-Act (PDSA), 211
Plan of Actions and Milestones (POA&M), 21
Plan reports, 20
PLAN-THINK-DRAW-WRITE, 203
Post-award debriefings, 191–92
Practical Project Management for Engineers (Patel), 144–45
Practice Standard for Project Risk Management, 144
Practice Standard for Work Breakdown Structures, 145
Preaward Survey of Prospective Contractor Accounting System, 238–39
Presenters, selecting, 92–93
Price data, 237
Price proposal reference, 236
Price to win (PTW), 234
Price websites and templates, 235

Pricing
cost objective and, 216–17
data, 233
direct costs and, 217
direct labor and, 217
fee and, 221
final cost objective and, 217
fringe benefits and, 219
fully burdened rate (FBR) and, 221
G&A costs, 220–21
indirect costs and, 218–19
labor categories (LCATs) and, 218
other direct costs (ODCs) and, 218
overhead (OH) and, 219–20
procurement costs and, 220
subcontractor labor and, 218
terminology, 216
unallowable costs and, 221–22
Win Strategy white paper and, 150
See also Cost and price volume
Prime contractors, 8, 248
Procurement costs, 220
The Program Management Office (Letavec), 141–42
"Progressive elaboration," 208
Project Management Body of Knowledge (PMBOK)
about, 18
in focusing on capture management, 20–22
framework, 18
INPUTS, 19, 21, 22, 24
lens, looking through, 18–20
OUTPUTS, 20, 22, 23–24, 25–26
"project" and, 201
TOOLS & TECHNIQUES, 20, 21, 22–23
Project management office (PMO), 15, 80
Project Management Professional (PMP) certification, 191
Proposal action items, in proposal directive, 112
Proposal audiences, 97–98
Proposal development
in a blanket, 14–16
box-in-a-box model, 137–39
communication in, 86–88
compliance and, 62
comprehensive risk management approach, 88–89
developing, 113
framing and, 83–85
future, reviews for, 41
information inclusion and, 84
international commercial, 244–46
knowledge management (KM) and, 39–48
as not a 9-to-5 position, 254
reactive, transitioning from, 94
remote environments, 123–24
SAM and Scrum view of, 202
Section M as a window and, 125–30
for successful Blue Team Review, 22–24
See also Solution development
Proposal development process
acronym definition in, 179–81
contact person and, 9
importance of, 76
priorities, 10
Proposal directive, 111–13
Proposal editing, checklist, 177–78
Proposaling
in context, 6–7, 13–26
counterbalancing demands of, 254
desire to win and, 61
font and line spacing and, 63–64
human interaction and, *xvii*
successful, 27
Proposal innovation centers (PICs), 30
Proposal integration map
about, 81
illustrated, 82
visualization and, 81, 82–83
Proposal journalists, 254–55
Proposal knowledge integrators (PKIs), 30
Proposal knowledge teams (PKTs), 30
Proposal manager
capture manager and, 97, 113
communication and, 94
critical role of, 113–16
goals, 115–16
responsibilities, 114
Proposal protest
source for, 162–63
source materials, 163–64
Proposal publications, 141–46
Proposal Readiness Work Products
about, 203
applying, 120
elevator speech, 203, 207
embedded in proposal document, 207
illustrated, 121–22
T1 Chart, 204, 205
T2 Chart, 204, 206, 207

Proposal response life cycle, 3
Proposals
  in context, 6–7
  getting started with, 253–57
  as knowledge-based documents, 167–68
  polishing, 208
  "responsible" contractor and, 89–90
  showing understanding in, 101–3
  successful, reusing and, 94–95
  successful versus failed, 98
Proposal sections, summarizing content, 174
Proposal support platform, 162–63
Proposal volume leads, 112
Proposal win
  bonuses, power of, 3
  bonus pool, 4–5
  what does it look like, 71–73
Proposal Win Party, 4
Proposal writing
  ABCs of, 168–70
  abstracting and, 171–72
  accuracy, 168–69
  acronyms, defining, 179–81
  action captions, 175–76
  active voice and, 172–75
  approach, 170
  brevity in, 169
  clarity in, 169
  contributors, 172
  conventions, 95
  dovetailing, 181–83
  guidance, 171
  humanizing, 96–97
  methods for enhancing, 176–77
  research, 177
  resources, 178
  sentence length, 170
  stages, 176–77
  standards, 170–71
  thinking graphically and, 175
  use of descriptive verbs, adjective, and adverbs, 173–74
  words to avoid in, 183–85
Publications, proposal, 141–46

Q

Quality Assurance Surveillance Plan (QASP), 89, 243
Quality management system (QMS), 190
Questions
  general, and answers, 9–11
  management approach (MA), 53–54
  technical approach (TA), 54–55
  training participants, 191
  unstated criteria, 53–55

R

Reasonableness, Section M and, 127
Recruiting process, 86
Red Team Review, 207, 208
Remote collaboration
  best practices, 123–24
  facilitator, 123
  humanizing, 124
Request for proposal (RFP)
  final, sharing, 55–56
  inconsistencies, 65–66
  release of, 2
  review, 226
Request for quotation (RFQ), 27
Requirements management, 15–16
Resource analysis, 173
Responsible, accountable, consulted, and informed (RACI) matrix, 173
Responsive web design (RWD), 162
Résumés, 118, 244
Reviewer's role, 8
Reviews
  for future proposal development, 41–42
  horizontal, 195–97
  perspectives, 195–211
  process, constructive an facilitated, 77
  RFP, 226
  "rolling review," 207
  SAM and Scrum view of, 202
  terms and acronyms for, 34
  vertical, 195–97
  See also specific types of reviews
Risk management and mitigation, 150
Risk Management Framework (RMF), 88–89
Risk management plan (RMP), 173
Rules of engagement, proposal meetings, 119–20

S

SADBUS (Small and Disadvantaged Business Utilization Specialist), 249–50
Salary ranges, determining, 227
Scaled Agile Framework (SAFe), 81–82

Schedule control, 77
Scrum
connect points, 201
proposal development and review and, 202
SAM and, 200–210
Scrum of Scrums (SoS) methodology, 210
Search Engine Optimization (SEO), 162–63
Section C, Statement of Work (SOW), 66
Section H, Special Contract Requirements, 66
Section L, Instructions to Offerors
compliance matrix and, 66
importance of understanding, 101–9
language, 55, 56, 57, 58
long-term business strategy, 57
process innovations, 58
proposal outline and, 182
Section M, Evaluation Factors for Award
completeness and, 127
compliance matrix, 66
comprehensiveness and, 129–30
corporate experience and, 129
creativity and, 128
effectiveness and, 127
efficiency and, 128
familiarity and, 129
feasibility and, 128
importance of understanding, 101–9
innovation and, 128
language, 55, 56–57, 58, 59
long-term business strategy, 58
merit and, 130
process innovations, 59
reasonableness and, 127
technical approach (TA) and, 120
technical expertise and, 129
using as a window, 125–30
Service level agreements (SLAs), 9, 54, 120
Significant strengths, 74–75
Situation, action, results (SAR) framework, 41–42
Small Business Development Centers (SBDCs), 250
Small businesses
federal subcontracting goals, 247
with PMO, 80
size standard, 164
SWOT analysis and, 80
winning proposals and, 79
Small Business Liaison Officers (SBLOs), 250
Small disadvantaged businesses (SDBs), 247
The Snowball Effect (Bounds), 143
Solution development
about, 15
brainstorming and, 139–46
information collection practice, 135–36
meeting participants, 134–35
overview, 133–34
process, 134–36
rule of engagement, 135
technical, framework for, 136–37
tools, 134
Win Strategy white paper as tool, 149–52
Source Evaluation Board (SEB), 88
Source selection decision statements (SSDS)
about, 24, 187
following proposal evaluation, 189
use of, 189
Source selection documents (SSDs)
about, 24, 187
proposal strengths and, 187–89
searches, 187–88
SSDSs, 24, 187, 189
SSSs, 192–94
Source selection statements (SSS)
about, 24
leveraging, 192–94
strengths in, 197–200
See also Source selection documents
Staffing plan, 148, 150
Stakeholders, 29, 53, 71, 107, 108
Stated criteria, 51–52
Statement of objectives (SOO), 226
Statement of work (SOW)
ability to perform, 226
approach to meeting requirements in, 243
"shall statements," 7
Strategic partnering, 247–49
Stratospheric Observatory for Infrared Astronomy (SOFIA), 139–40, 141
Strength(s)
achieving through forward-looking business decisions, 187–94
candidate, percentage of, 199–200
discriminators and, 97
"evidence of," 199

going beyond expectations and, 98
"jumping off the page," 198
linking business decisions to, 190–91
as proposal aspect, 198
in proposal directive, 112
significant, example, 192, 193
source selection documents and, 187–89
in source selection statement, 197–200
understanding and, 107
Strengths, weaknesses, opportunities, threats (SWOT) analysis, 21, 80, 151
Subcontractor labor, 218
Subcontractor management, 150
Subject matter experts (SMEs)
acronym definitions and, 180
brain-writing, 141
core competencies, 116
interviewing, 114, 116–19
"pulling intriguing threads" and, 117
questions for, 116–17
résumés for, 118
role, 8
Successive Approximation Model (SAM)
combining Scrum, 200–210
connect points, 201
for instructional design (ID), 201
modified, 208
proposal development and review and, 202
proposal life-cycle and, 201–3
Supplier Diversity Programs, 250
Supporting information, 230–31
System for Award Management (SAM), 61–62

T

Tables, completeness, reasonableness, effectiveness, efficiency, 128
Task Order Request for Proposal (TORFP), 164
Technical approach (TA)
about, 150
aligning cost proposal with, 226–27
elements, 82
questions, 54–55
sections, building, 120–23
Win Strategy white paper and, 148
Technical expertise, Section M and, 129
Technical interchange meetings (TIMs), 5, 17
Technical Solution (TS), 136–37
Technology Readiness Levels (TRLs), 204
Templates
price/cost, 235
using, 94
The 3-Minute Rule (Pinvidic), 145
Timeboxing, 209
Time management, 77
TOOLS & TECHNIQUES
in business development, 20
for capture management, 21
in knowledge management (KM), 24–25
for successful Blue Team Review, 22–23
To Sell is Human (Pink), 143
Total compensation plan (TCP), 134
Total contract value (TCV), 4–5
Total cost input (TCI), 221
Total cost of ownership (TCO), 245
Training/cross-training process, 86
Triple bottom line (TBL), 245

U

Unallowable costs, 221–22
Uncompensated overtime, 231
Understanding
approaching, 102
examples of, 104
missing the point of, 104–9
section, example, 105–6
sections, beginning, 104–5
showing in proposals, 101–3
strength and, 107
taking it seriously, 107
"we understand" and, 103
Uniform contract format (UCF), 125–30, 245
Unique Entity Identifier (UEI), 155
Unstated criteria
about, 51–52
evaluation, 52–53
management approach questions, 53–54
questions to learn about, 53–55
technical approach questions, 54–55
See also Criteria

V

Value Stream Map (VSM), 24
Vertical reviews, 195–97
Veteran-owned businesses, 247
V formation, 30–31

Virtual teams
high-performance, 87–88
potential risks, mitigating, 87
success, communications in determining, 86
trust in, 86, 88
Visual communication. *See* Graphics
Visualization, 43–44
Volatile, uncertain, complex, and ambiguous (VUCA), 26, 29, 35

W

Websites
price/cost, 235
in proposal development, 256
proposal support platform, 162–63
"What's in it for me?" (WIFM), 72
Winning proposals
business acquisition as formal process and, 78–80
contingency factors, 95
"good is the enemy of great" and, 73–75
most important factors in, 75–78
small businesses and, 79
what do they look like, 71–73
WIFM question and, 72
Win Strategy white paper
about, 147–48
approach to building, 148–49
as guidepost, 149
length, 152
as solution development tool, 149–52
Women-owned businesses (WOBs), 247
Words to avoid, 183–85
Workload fluctuations, 15
Workshare percentage, 9
Writing
abstracting and, 171–72
active voice and, 172–75
competitive federal, 243–44
grant proposal, 242–43
guidance, 171
standards, 170–71
See also Proposal writing

Artech House Technology Management and Professional Development Library

B. Michael Aucoin, Series Editor

Actionable Strategies Through Integrated Performance, Process, Project, and Risk Management, Stephen S. Bonham

Advanced Systems Thinking, Engineering, and Management, Derek K. Hitchins

Applying Total Quality Management to Systems Engineering, Joe Kasser

Building Successful Virtual Teams, Francine Gignac

Critical Chain Project Management, Third Edition, Lawrence P. Leach

Decision Making for Technology Executives: Using Multiple Perspectives to Improve Performance, Harold A. Linstone

Designing the Networked Enterprise, Igor Hawryszkiewycz

Electrical Product Compliance and Safety Engineering, Volume 2 Steli Loznen and Constantin Bolintineanu

Engineer's and Manager's Guide to Winning Proposals, Donald Helgeson

Engineering and Technology Management Tools and Applications, B. S. Dhillon

Enterprise Release Management: Agile Delivery of a Strategic Change Portfolio, Louis Taborda

The Entrepreneurial Engineer: Starting Your Own High-Tech Company, R. Wayne Fields

Evaluation of R&D Processes: Effectiveness Through Measurements, Lynn W. Ellis

From Engineer to Manager: Mastering the Transition, Second Edition, B. Michael Aucoin

Global High-Tech Marketing: An Introduction for Technical Managers and Engineers, Jules Kadish

How to Become an IT Architect, Cristian Bojinca

Integrated IT Project Management, Kenneth R. Bainey

Introduction to Information-Based High-Tech Services, Eric Viardot

Introduction to Innovation and Technology Transfer, Ian Cooke and Paul Mayes

ISO 9001:2000 Quality Management System Design, Jay Schlickman

IT Project Portfolio Management, Stephen S. Bonham

Managing Complex Technical Projects: A Systems Engineering Approach, R. Ian Faulconbridge and Michael J. Ryan

Managing Successful High-Tech Product Introduction, Brian P. Senese

Managing Virtual Teams: Practical Techniques for High-Technology Project Managers, Martha Haywood

Mastering Technical Sales: The Sales Engineer's Handbook, Fourth Edition, John Care

The New High-Tech Manager: Six Rules for Success in Changing Times, Kenneth Durham and Bruce Kennedy

The Parameter Space Investigation Method Toolkit, Roman Statnikov and Alexander Statnikov

Planning and Design for High-Tech Web-Based Training, David E. Stone and Constance L. Koskinen

A Practical Guide to Managing Information Security, Steve Purser

Practical Model-Based Systems Engineering, Jose L. Fernandez and Carlos Hernandez

Practical Reliability Data Analysis for Non-Reliability Engineers, Darcy Brooker with Mark Gerrand

The Project Management Communications Toolkit, Second Edition, Carl Pritchard

Project Managment Process Improvement, Robert K. Wysocki

Reengineering Yourself and Your Company: From Engineer to Manager to Leader, Howard Eisner

The Requirements Engineering Handbook, Ralph R. Young

Running the Successful Hi-Tech Project Office, Eduardo Miranda

Successful Marketing Strategy for High-Tech Firms, Second Edition, Eric Viardot

Successful Proposal Strategies for Small Businesses: Using Knowledge Management to Win Government, Private Sector, and International Contracts, Sixth Edition, Robert S. Frey

Successful Proposal Strategies On-the-Go!, Robert S. Frey

Systems Approach to Engineering Design, Peter H. Sydenham

Systems Engineering Principles and Practice, H. Robert Westerman

Systems Reliability and Failure Prevention, Herbert Hecht

Team Development for High-Tech Project Managers, James Williams